Modern Development and Challenges in Virtual Reality

Edited by Mamata Rath
and Tushar Kanta Samal

Published in London, United Kingdom

Modern Development and Challenges in Virtual Reality
http://dx.doi.org/10.5772/intechopen.104075
Edited by Mamata Rath and Tushar Kanta Samal

Contributors
Jufan Zhang, Yao Zhou, Fengzhou Fang, Trung Hieu Tran, Yirui Jiang, Leon Williams, Ingrid Winkler, Camila Rossi, Marinilda Lima, Alex Álisson Santos, Pavel M. M Trivailo, Hirohisa Kojima, Rares-Catalin Catalin Nacu, Daniel Fodorean, Carlos Augusto Rengifo Espinosa, Oton Alberto Navas de la Cruz, Pierre Boulanger, Thea Wang, Mahdi Rahmani Hanzaki, Dominique Michaud, Geoffrey Edwards, Jocelyne Kiss, Jonathan Proulx-Guimond, Jonathan Caron-Roberge, Ernesto Morales, Cyril Schneider, Mary Thaler

First published in London, United Kingdom, 2023 by IntechOpen
IntechOpen is the global imprint of INTECHOPEN LIMITED, registered in England and Wales, registration number: 11086078, 5 Princes Gate Court, London, SW7 2QJ, United Kingdom

British Library Cataloguing-in-Publication Data
A catalogue record for this book is available from the British Library

Additional hard and PDF copies can be obtained from orders@intechopen.com

Modern Development and Challenges in Virtual Reality
Edited by Mamata Rath and Tushar Kanta Samal
p. cm.
Print ISBN 978-1-83768-432-8
Online ISBN 978-1-83768-433-5
eBook (PDF) ISBN 978-1-83768-434-2

Meet the editors

Dr. Mamata Rath obtained her Ph.D. in Computer Science from Siksha 'O' Anusandhan (SOA) University, Odisha, India. Her academic interests include subjects of computer science and information technology. Currently, she is a professor in the Department of Computer Science and Engineering, Dhaneswar Rath Institute of Engineering and Management Studies (DRIEMS) Autonomous Engineering College, Cuttack, Odisha, India. Dr. Rath has been an organizing member and chair for numerous international conferences. She has published forty-eight research articles in reputed journals and international conferences and edited four books on artificial intelligence, machine learning, and deep learning. She is a regular reviewer of research papers and was the program chair of the 2020 International Conference on Advanced Computing and Intelligent Engineering (ICACIE).

Dr. Tushar Kanta Samal obtained a BTech in Information Technology from Biju Pattanaik University of Technology, Rourkela, Odisha, India. He obtained an MTech and Ph.D. in Computer Science and Engineering from Veer Surendra Sai University of Technology, Odisha, India. He has more than 10 years of teaching and research experience. At present, he is a senior assistant professor in the Department of Computer Science and Engineering, Dhaneswar Rath Institute of Engineering and Management Studies (DRIEMS) Autonomous Engineering College, Cuttack Odisha, India. His research interests include wireless sensor networks, wireless body area networks, and cloud computing.

Contents

Preface IX

Chapter 1 1
Advancements in Optical See-through Near-Eye Display
by Jufan Zhang, Yao Zhou and Fengzhou Fang

Chapter 2 25
Smart Garment Design for an Augmented Reality Body Mapping Experience
by Dominique Michaud, Geoffreyjen Edwards, Jocelyne Kiss, Jonathan Proulx-Guimond, Jonathan Caron-Roberge, Ernesto Morales, Mary Thaler and Cyril Schneider

Chapter 3 43
Virtual Reality Utilization in Electrical Vehicle Development
by Rares-Catalin Nacu and Daniel Fodorean

Chapter 4 67
Applications of Mixed Reality for Smart Aviation Industry: Opportunities and Challenges
by Trung Hieu Tran, Yirui Jiang and Leon Williams

Chapter 5 81
Learning Composition and Architectural Design with Immersive Virtual Reality Application CREALITY 1.0
by Carlos Augusto Rengifo Espinosa and Oton Alberto Navas de la Cruz

Chapter 6 91
A Participatory Content Authoring Workflow for Augmented Reality at Industrial Maintenance
by Camila Rossi, Marinilda Lima, Alex Álisson Santos and Ingrid Winkler

Chapter 7 109
Improving Medical Simulation Using Virtual Reality Augmented by Haptic Proxy
by Pierre Boulanger, Thea Wang and Mahdi Rahmani Hanzaki

Chapter 8 133
Conjugated 3D Virtual Reality Worlds in Spacecraft Attitude Control
by Pavel M. Trivailo and Hirohisa Kojima

Preface

Modern Development and Challenges in Virtual Reality focuses on the advanced concepts of virtual reality (VR), an environment that gives the user the impression that they are completely engrossed in their surroundings. A virtual reality headset, helmet, or other equipment is used to view this environment. VR enables us to, for example, learn how to conduct heart surgery, improve our sports training, and immerse ourselves in video games as if we are one of the characters. One of the technologies with the highest expected future growth is VR performance. Applications that go beyond leisure, tourism, or marketing are now in high demand on the market, and thus they must be user-friendly and economical.

Chapter 1, "Advancements in Optical See-through Near-Eye Display", by Jufan Zhang, Yao Zhou and Fengzhou Fang describes the development of the optical see-through near-eye display, which can be applied in medical devices as well as in the fields of education, aviation, entertainment, and more. Typical products include head-mounted displays (HMDs) and augmented reality (AR) glasses. The optical display system of AR devices consists of a miniature projecting module and an optical imaging module. This chapter summarizes the display systems used by AR glasses on the market, including various mini-display screens and optical imaging elements. The differences in optical combinators are a key factor for distinguishing various AR display systems. Thus, it is essential to determine the advantages and disadvantages of each optical imaging technology applied in this area. In addition, the characteristics of the projectors are crucial to the quality of the images.

Chapter 2, "Smart Garment Design for an Augmented Reality Body Mapping Experience", by Dominique Michaud et al., describes the development of wide handmade belts as supports for a live AR event. It describes how the garments were made to best meet the constraints of the AR design and how they were used as part of a fashion show and book promotion event. The authors determined the types of features the AR software can detect and then created prototypes based on those characteristics. The belt design followed several steps: (1) computer image preparation, (2) thread preparation, (3) thread dyeing, (4) weaving, (5) assembly, and (6) photographing the result and retouching the image. The lack of adjustability to deformation was a concern when dealing with clothing and fabric. The authors resolved this issue by stiffening the fabric. The authors also discuss the performance of the belts during the live event, other textile techniques as substitutes (e.g., patchwork), and future research prospects.

Chapter 3, "Virtual Reality Utilization in Electrical Vehicle Development", by Rares-Catalin Nacu and Daniel Fodorean, describes a new perspective on vehicle testing methods by using driving simulators, which can be employed starting from the early phases when components are designed to the real manufacturing stage. Based on these driving simulators, replicating real driving conditions can be a complex task, and the vehicle components can be dynamized with data coming from it. Therefore, buildings, traffic signs, road types, and vehicles can be imitated using a VR environment and inspired by the real world or the imagination but conclusive for the tested components. Moreover, for more realistic tests, a human driver can be included in the VR environment to control the virtual vehicle, leading to more reliable results. The chapter discusses many types of simulators but focuses on a specific type that is capable of testing the vehicle's propulsion system and its driver assistance systems. The chapter ends with a case study in which different configurations, software, and hardware are tested.

Chapter 4, "Applications of Mixed Reality for Smart Aviation Industry: Opportunities and Challenges", by Dr. Trung Hieu Tran, Yirui Jiang and Leon Williams, suggests that mixed reality has improved operational efficiency and enhanced passenger experience in the aviation industry. Integrated with advanced machine learning and artificial intelligence techniques, mixed reality can easily deal with tons of aviation data to support decision-making processes in this industry. The chapter presents the state-of-the-art applications of mixed reality in the smart aviation industry. It discusses the opportunities and challenges of integrating mixed reality with advanced machine learning and artificial intelligence techniques into the aviation industry. The chapter focuses on how integrated mixed reality can improve the quality and reliability of maintenance, operation, piloting, training, and product design in smart aerospace engineering. It also describes autonomous, self-service, and data visualization systems in smart airports to enhance passenger experience. Finally, the chapter discusses airlines' digital-based responses to the COVID-19 crisis.

Chapter 5, "Learning Composition and Architectural Design with Immersive Virtual Reality Application CREALITY 1.0", by Carlos Augusto Rengifo Espinosa and Oton Alberto Navas de la Cruz, focuses on identifying how the CREALITY 1.0 application (self-development) played an important role in teaching and teaching processes in architecture, working with topics such as architectural composition and supported by immersion in VRs that could stimulate student creativity in a college education setting. As a theoretical basis, this research and exploration exercise focuses on the composition and formal principles underpinning student-generated design proposals. The evaluative component of the exercise is directed toward the analysis of evidence of learning using the application, and it was possible to emphasize that the advantages in execution time, expenses of materials and dynamization in the cognitive processes allow considering the application as a relevant tool with the field of education in architecture and the present world.

Chapter 6, "A Participatory Content Authoring Workflow for Augmented Reality at Industrial Maintenance" by Camila Rossi, Marinilda Lima, Alex Álisson Santos and Ingrid Winkler explains that the adoption of AR-based instructions enhances maintenance operations by shortening job completion time and reducing errors. However, scaling AR in industrial settings remains costly since content authoring demands computational skills such as 3D modeling and programming. Furthermore, processes can easily become obsolete, causing maintainers to abandon written instructions. Therefore, the chapter proposes an AR-based participatory content authoring workflow for maintenance tasks. The authors followed the Design Science Research paradigm, including a literature review, conception of a workflow, and simulation to evaluate the proposed workflow's validity. They found that current workflows overlook participatory content authoring involving maintainers and that most research focuses on describing the technical architecture of proposed systems rather than a workflow that supports the use of technology in industrial settings. Regarding the proposed participatory workflow, most respondents stated it was simple to use, improved their capacity to develop AR content, and would help the industry adopt AR. As a result, the participatory authoring workflow can optimize AR content authoring during maintenance, encouraging the maintainers' interaction, and providing opportunities for procedure improvement. The chapter concludes that non-programmer-friendly AR software tools save content production time while enhancing users' perceptions of their own technological talents.

Chapter 7, "Improving Medical Simulation Using Virtual Reality Augmented by Haptic Proxy", by Dr. Pierre Boulanger, Thea Wang and Mahdi Rahmani Hanzaki, explores how the realism of haptic perception in VR can be significantly enhanced with the help of the concept of haptic proxy. In haptic proxy, the position and orientation of physical objects are tracked in real-time and registered to their virtual counterparts. A compelling sense of tactile immersion can be achieved if the tracked objects have similar tactile properties to their virtual counterpart. A haptic proxy prototype was developed and a pilot study was conducted to determine if the haptic proxy system is more credible than standard VR. To test the prototype, the author performed simple medical tasks such as moving a patient's arm and aiming a syringe at specific locations. Results suggest that simulation using a haptic proxy system is more believable and user-friendly and can be extended to developing new generations of open-surgery simulators.

Chapter 8, "Conjugated 3D Virtual Reality Worlds in Spacecraft Attitude Control", by Prof. Pavel M. Trivailo and Hirohisa Kojima, describes the application of VR possessing inertial morphing (IM) capabilities in spacecraft movement control. The concept of IM proposed/applied by the researcher(s) in 2017, is attractive for the design of fully autonomous future space missions, as swift control, requiring minimal energy and computations, can be achieved with exiguous/paltry morphings. To assist rapid planning of the optimized maneuver scenarios, the chapter proposes collocated merging of the various VR simulation worlds for the same spacecraft or merging of the VR worlds for the same system (but in different configurations). This enables concurrent utilization of different methods of modeling, including Poinsot's and Bine's construction and Euler's equations. Therefore, superimposed VR worlds are called "conjugated." The chapter presents the classical methods of modeling torque-free systems and demonstrates their implementation in the developed VR-integrated, interactive package. The effectiveness of the VR-conjugated environment is illustrated by its use for planning of spacecraft de-tumbling and 180- and 90-degree inversions. The developed VR environment enables utilization of both the body-axes spacecraft coordinate system and inertial coordinate system with instant transition from one into another, switching on/off various VR worlds for multiple supports in the process of mission design.

Dr. Mamata Rath
Professor (Computer Science),
Driems(Autonomous) Engineering College,
Cuttack, Odisha, India

Dr. Tushar Kanta Samal
Senior Asst. Professor (Computer Science),
Driems(Autonomous) Engineering College,
Cuttack, Odisha, India

Chapter 1

Advancements in Optical See-through Near-Eye Display

Jufan Zhang, Yao Zhou and Fengzhou Fang

Abstract

With the development of optical design and manufacturing, the optical see-through near-eye display becomes a promising research topic in recent decades, which can be applied in medical devices, education, aviation, entertainment et al. Typical products include Head-mounted Displays (HMDs) and Augmented Reality (AR) glasses. The optical display system of AR devices mainly consists of a miniature projecting module and an optical imaging module. In this chapter, the display systems used by AR glasses on the market, including various mini-display screens and optical imaging elements, have been systematically summarized. Therein, the differences in optical combinators are the key part to distinguish various AR display systems. Thus, it is essential to figure out the advantages and disadvantages of each optical imaging technology applied in this area. Besides, the characteristics of the projectors are crucial to the quality of the images.

Keywords: see-through near-eye display, augmented reality, optics, metaverse, projector, freeform optics, waveguide, off-axis optics, birdbath optics

1. Introduction

The optical see-through near-eye display is a promising solution for various industrial sectors, such as education, entertainment, military, tourism, etc. Typical products include Head-mounted Displays (HMDs) and Augmented Reality (AR) glasses. Since Facebook announced the Metaverse, more and more companies released their products. This technology is a medium and continuum that spans between the real world and the virtual world [1–3], by integrating the real environment with the projected virtual objects. Technically, the optical module combines the imaging system and the projecting system, as shown in **Figure 1**, both of which directly influence the optical performance and imaging quality. The light from the projector is transferred by the imaging optics without blocking the view of the real world.

Over the past decades, there is a significant advancement in the see-through near-eye display. Although there are a number of technical drawbacks and problems limiting its success in the consumer market, the achievements made in the enterprise market have witnessed value and high commercial potential. The latest technological progress brings us new possibilities for a wider range of applications in the near future [4–7]. This chapter focuses on the current mainstream optical imaging technologies

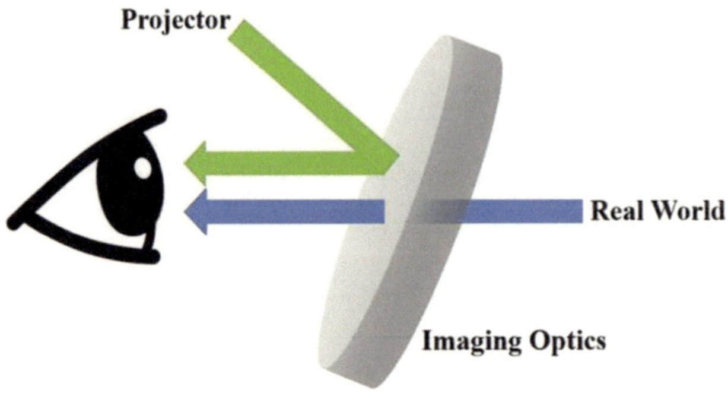

Figure 1.
Schematic diagram of the optical see through near-eye display.

and projecting technologies, along with typical applications, to offer the readers a general idea on the state of the art.

1.1 Applications

With the maturity of AR technology, AR is increasingly used in various industries, such as education, training, medical care, design, advertising, etc. AR has injected new vitality into education with its rich interactivity. Compared with stuffy paper books, AR combines text and dynamic 3D images together, which provides an immersive feeling and facilitates a quicker understanding of the information. AR enhances the clarity and intuitiveness and perceptual impact of real situations, making situational learning more friendly, dynamic, and natural [8–11]. E.g., for training on medical and clinic treatment, the application of AR makes the esoteric and profound medical theories livelier and more concrete, which greatly improves teaching efficiency and quality, especially in minimally invasive surgery [12–15]. In tourism, such as museums, AR provides virtual text, pictures, videos, and other information for the introduction and description of exhibits. And AR has also been applied for restoring and displaying cultural relics by virtually filling up the incomplete part, which brings an immersive feeling to tourists [16, 17]. In addition, in the industry, based on AR smart glasses, data collection and processing are carried out through the AR cloud to provide visual information for the technical support team, therefore realizing expert-level remote assistance. This truly makes the communication of industrial issues more direct, accurate, efficient through visualized way, and eliminates the risk of unavailability of qualified experts in the field for urgent problems, and also significantly saves cost and time. AR helps abandon complicated work manuals, flowcharts, walkie-talkies, etc., and completely liberates the hands of workers/operators [18]. In the military, pilots can use HMD to observe navigation information and even crucial information about the enemy. The Synthetic Training Environment based on AR systems helps soldiers be trained in a more immersive way, by placing them in a more physically and psychologically stressful combat environment [19–21]. In the entertainment industry, AR is applied to create interactive games, like racetracks, and basketball fields, for which the camera can track the locations or even the body language to give a more accurate response [22–24].

Among the above application fields, the most commonly used hardware are head-mounted displays (HMDs) and handheld devices like AR glasses. HMD is worn on the head like a see-through helmet, which provides images in front of the eyes. An HMD can facilitate many uses including aviation, gaming, medicine, and engineering. AR glasses are worn like regular glasses in front of the eye, but function in a different way. Smart AR glasses are computer-enabled wearable devices that add virtual information to a user's real-world scene by superimposing computer-generated or digital information, such as 3D images, animations and videos. The information can be retrieved from PC, phones, or other devices, which can be supported by Wi-Fi, mobile data, or Bluetooth.

1.2 Key optical parameters

The optical performance is closely affected by the specific parameters of the display system. Rather than a single parameter determining the optical performance, there is a trade-off among the following major parameters.

Field of view It refers to the solid angle between the outline of the object observed by the human eye and the line connecting the center of the pupil of the human eye. FOV includes the vertical field of view, horizontal field of view, and diagonal field of view [25]. Normally, the size of a person's retina is limited, so the corresponding viewing range of the human eye is also limited. FOV is the major indicator that many developers are concerned with primarily.

Eye relief Eye relief of an optical display represents the distance from the exit pupil area to the optical combiner within which the user's eye can obtain a full view and clear images. Most near-eye displays need binoculars or monocular with a minimum of 16 mm eye relief [26]. It is an important parameter to comfort the use.

Eye box Eye box is defined by the space within which the images can still be effectively viewable. So, images within the eye relief can be observed in both angular and lateral movements of the human eye. It's how far off center your eye can be and still see through the scope properly. Within the eye box area, the observer at any position can reach the entire FOV. Exceeding this area may result in distorted images, incorrect color rendering, or even no content [27]. A larger eye box allows the user to have greater freedom and head movement to observe the whole visual image.

Chromatic aberration Chromatic aberration refers to the phenomenon that optical lenses cannot focus all wavelengths of colored light on the same point. Chromatic aberration is caused by the phenomenon of lens scattering. It may appear as blurred or obvious color fringing around objects in the image, especially in the case of high contrast. Only one point will be focused by a perfect lens with the smallest circle of a blur. According to the wavelength dispersion of different planes, chromatic aberration can be divided into two types: longitudinal chromatic aberration and lateral chromatic aberration [28–30]. The optical element in the combiner may cause chromatic aberration, which affects the optical performance and imaging quality.

Distortion Lens distortion is actually a general term for the inherent perspective distortion of optical lenses, that is, the distortion caused by perspective. There are three kinds of distortions including pincushion distortion, barrel distortion, and linear distortion. Pincushion distortion is a phenomenon caused by the lens "shrinking" the picture toward the center. Barrel distortion is a barrel-shaped expansion caused by the physical properties of the lens and the combined structure of the lens. Linear distortion is defined as a change in amplitude or phase with no new frequencies added [31–33].

Stray light For the imaging optical system, any undesired light that propagates to the detector face can be defined as stray light. Due to the multiple optical elements and complex architecture used in an integrated display system, stray light may be caused by diffraction, unwanted reflection, and scattering. In a sense, stray light has a veto effect on optical systems. If the stray light affects the imaging quality, all ray paths should be traced back to the receiver for defects shooting. However, stray light cannot be completely eliminated, but can only be suppressed to a certain extent [34, 35]. As long as stray light is controlled within an imperceptible range by human eyes, or within some acceptable or permissible extent defined by users, it is regarded as the completion of stray light suppression.

Brightness and transmittance Brightness refers to the amount of light in the virtual image displayed by the optical system. Enough brightness allows you to see the image clearly in direct sunlight. It is also one of the major challenges faced by current AR headsets. The brightness of most existing AR glasses can be only used indoors, and is almost unusable outdoors. In order to alleviate this problem, some AR headsets use optical designs such as birdbath to block ambient light or use tinted lenses to improve the relative brightness of the optical module, but the associated adverse effect is reducing the light transmittance of the optical module. Light transmittance refers to how much ambient light the human eye can receive through the optical element [36–39]. The ideal light transmittance is 100%. Of course, it is still difficult for the existing AR optical technology to achieve this. Low light transmittance may be acceptable to consumers for some specific application scenarios, but it is generally unacceptable in many professional/industrial application scenarios because light transmittance has a great impact on job safety.

Resolution and contrast Resolution refers to the number of pixels a display can cover, and the optimal display resolution should be close to or slightly beyond the limits of human vision. There's no official definition or measurement of contrast or contrast ratio, but to most people, it's a perception, simply a display's ability to produce both bright and dark pixels. With low contrast, bright content and dark content will not be displayed correctly. In the optical perspective AR display system, dark or black color is hard to render, so high transparency areas may appear dark color in low contrast. In short, the brighter the AR display, the higher the requirement for contrast. For AR display, the color perceived by human eyes is also related to the real environment background superimposed by the virtual image. As with contrast, pixels vary in color quality depending on where they are on the display [40, 41]. For example, the same pixel color may look different on the left and right sides of the display with distinct image patterns, as well as depending on the location of the user's pupils.

Vergence accommodation Human perception of a three-dimensional environment can be divided into psychological perception and physiological perception. Psychological perception includes five aspects of visual suggestion, such as shadow, occlusion, light, affine and texture, and prior knowledge. The physiological perception of stereoscopic vision mainly includes blurred focus, moving parallax, and binocular parallax. There are two main reasons for vertigo: [1] conflicts between binocular parallax and focusing blur on visual perception; [2] the conflict between motion perception and visual perception [6, 42–46].

Size and weight Size is one of the biggest challenges of the see-through near-eye display. Larger FOV and eye box always mean bigger size and weight at the same time. The larger size usually implies inconvenience to wear and tends to block sight. Besides, there is a limit to the amount of weight that the human ear, bridge of the nose, and top of the head can hold.

Delay All virtual images are produced by the projectors; thus, the response time of hardware is very important for the reaction of the human brain and eye. An imaging delay of fewer than 5 milliseconds is generally considered sufficient for optical perspective systems. A longer delay would cause dizziness.

2. Optical imaging technology

The optical display system of see-through near-eye devices is usually composed of micro-projector and imaging components. In general, the display system of AR glasses on the market at present is a combination of various micro-projectors and imaging components such as prisms, free-form surfaces, Birdbath, optical waveguides, etc. The difference in optical combination is the key part to distinguish the display systems.

2.1 Conventional optics

2.1.1 Off-axis optics/birdbath optics

More HMD optical systems have applied folding/reflecting structures with the high demands of modern air combat. The structure can meet the technical requirements of large eye relief, large exit pupil diameter, and large FOV. The system consists of two basic types: coaxial (rotational symmetric geometry) and off-axis (rotational asymmetric geometry). Compared with the coaxial system, the off-axis system can reach a wider FOV. Therefore, most HMD optical systems adopt the off-axis display [47, 48]. An example of the off-axis structure is illustrated in **Figure 1**, which reaches 40° × 30° FOV, 15 mm exit pupil diameter, 26.4 mm focal length and 25 mm eye relief. However, off-axis optical system usually contains many optical elements, which means larger size and heavier weight, so not suitable for long time use.

Another conventional technology is the birdbath structure. The polarized beam splitter (PBS) is used to split the light beam from the light source to the human eyes, which can be a cube or a film. Most products are applying this technology currently, such as google glass and Lenovo headsets, as shown in **Figure 2**. **Figure 2a** shows how google glasses work, which mainly includes a projector and a beam splitter (BS) prism. When the projected light travels to the BS prism, part of it is reflected to human eyes and produces virtual images, without blocking the light from the real world. **Figure 2b** shows

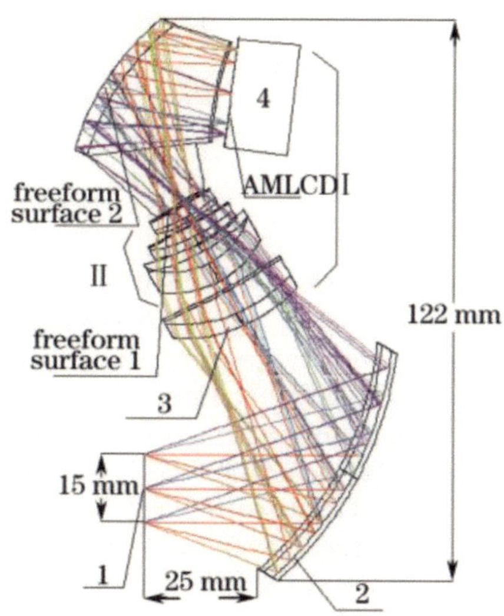

Figure 2.
Optical system of off-axis system [44].

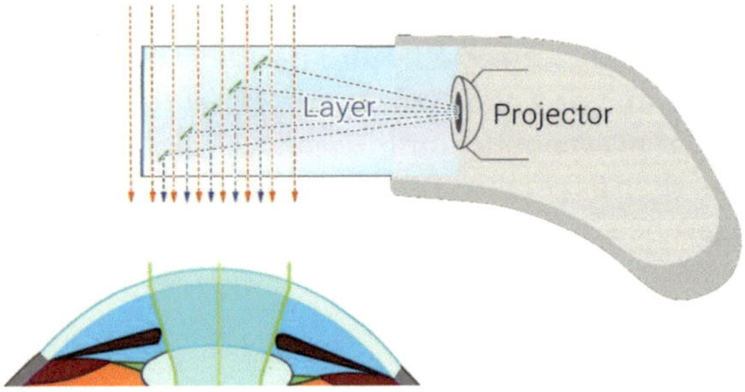

(a)

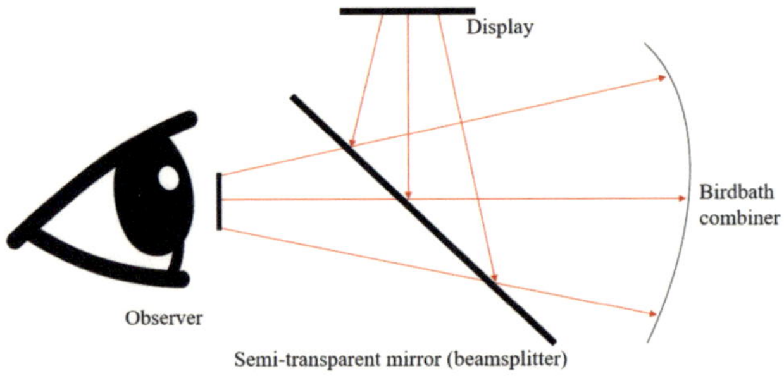

(b)

Figure 3.
Example of the birdbath structure in the see-through near-eye display. (a) The schematic diagram of Google Glass [51]. (b) The schematic diagram of Lenovo headsets.

how Lenovo headset works. The structure applies a plate beam splitter that allows light from a top-mounted display facing the bottom to pass through it and is reflected by a spherical mirror to the other side of the BS, to the viewer's eyes. The birdbath design can achieve larger FOV and larger eye relief, meanwhile, larger sizes are needed compared with off-axis technology [49–51]. As shown in **Figure 3**, the design space is defined by the thickness as a function of the eye box size. The window limits the minimum and maximum values on the eye box [52]. When applying the birdbath technology, the size and weight become proportional to the eye box and FOV, which will also constrain some other optical parameters (**Figure 4**).

2.1.2 Freeform optics

Freeform optics is defined as any non-rotationally symmetric surface or micro-array surface, which is very different from spherical and aspherical geometry. Freeform lenses can enable unique optical performance, such as low f-number, large eye relief, and wide FOV. One possible form of freeform optics is the eccentric use of rotating symmetrical lens, thus accommodating off-axis ranges. There are three main ways to describe free-form surfaces, including NURBS, XY polynomials, and radial basis function representation. The use of freeform optics reduces the number of components in an optical system, resulting in a smaller, lighter, and more efficient system [53–56]. The biggest advantage of freeform surfaces is that they can achieve very good imaging quality due to their special geometry and a high degree of freedom, as shown in **Figure 5**. However, compared with the traditional symmetric lens, the optical design and manufacturing process are more complex and demanding [57–60].

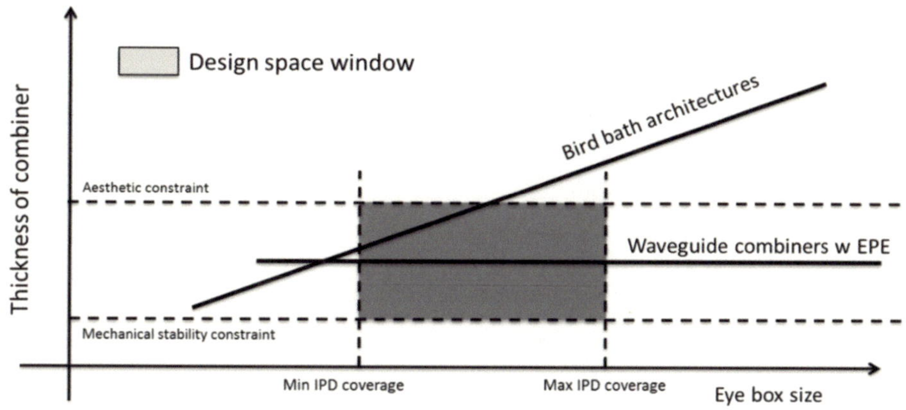

Figure 4.
Typical design space for specific interpupillary distance (IPD) coverage and ID requirements [52].

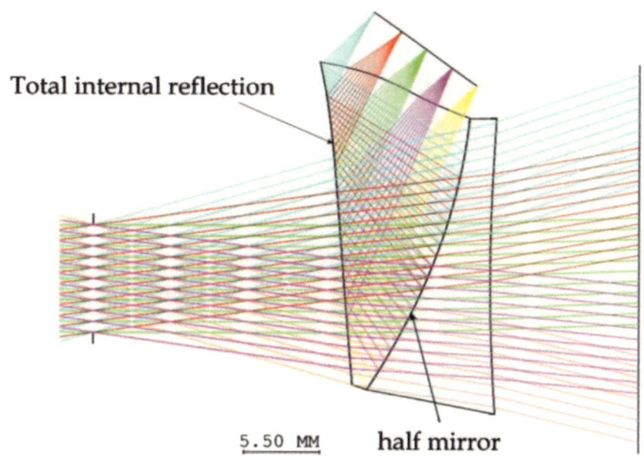

Figure 5.
Layout of the see-through HMD by coupling the FFS prism lens [53].

2.2 Waveguide optics

Optical waveguide technology is a distinctive optical component developed to meet the demands of the see-through near-eye display. Light travels at different speeds in different substances. Thus, when the light goes from one substance to another, it refracts and reflects at the interface of the two substances. Moreover, the angle of the refracted light changes with the angle of the incident light. When the incident angle reaches or exceeds a critical angle, the incident light will be reflected back without refraction, which is called the total internal reflection (TIR) of light. According to the principle of TIR in geometric optics, the light produces TIR at the interface between the waveguide and the air and forms the necessary conditions for the light to be bounced inside the waveguide and propagated forward without exiting the waveguide, as shown in **Figure 6** [61]. In virtue of the waveguide transmission mode, the display and imaging system can be moved away to the top of the forehead or the ear side of observer, which greatly reduces the obstruction of the optical system to the outside view, and makes the weight distribution more ergonomic, thus improving the wearing experience of the device [62–64]. The optical waveguide can

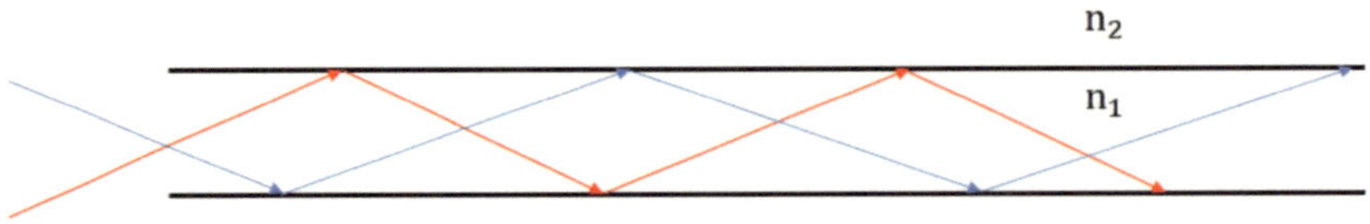

Figure 6.
The layout of the waveguide [61].

be divided into two types, the geometric waveguide and the diffraction waveguide, mainly based on the coupling structure of light in and out of the waveguide.

2.2.1 Diffractive waveguide

Diffractive optical waveguide includes surface relief grating (SRG) waveguide fabricated by lithography technology and volume holographic grating (VHG) waveguide fabricated by holographic interference technology, as shown in **Figure 7**. For the diffractive waveguide, the light source includes light emitting diode (LED), organic light emitting diode (OLED), liquid crystal on silicon (LCOS) and laser scanning display (LSD). The light from the light source is collimated to the waveguide in one direction. The in-coupler diffractive optical element (DOE) modulates the light propagation within the waveguide by TIR. When the light travels to the out-coupler, the condition of TIR is broken and the light is coupled out to human eyes. The virtual images

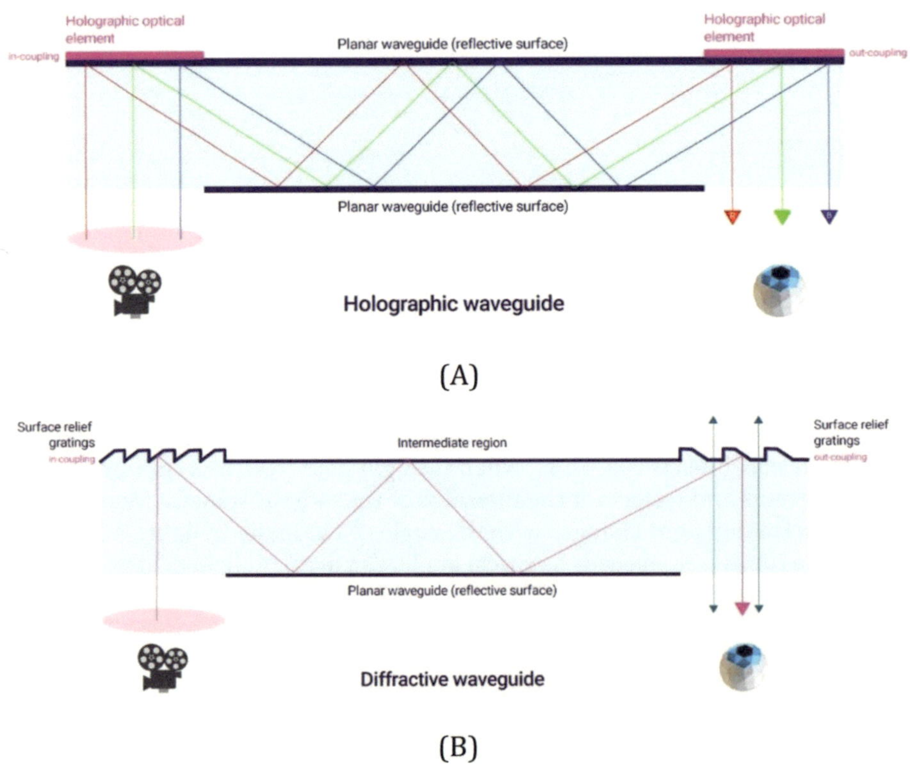

Figure 7.
The schematic of the diffractive waveguides [69]. (A) The schematic diagram of the SRG waveguide. (B) The schematic diagram of the VHG waveguide.

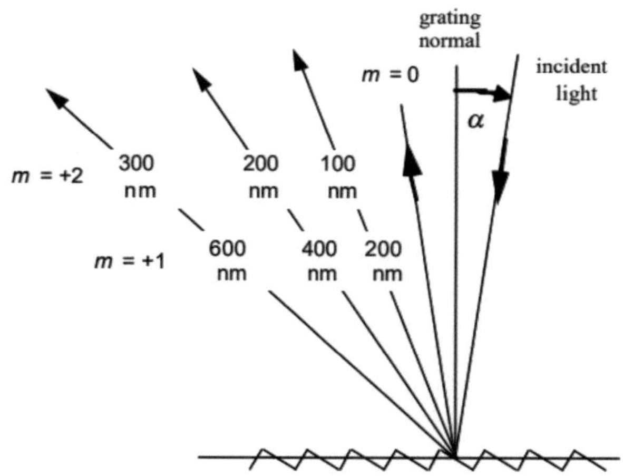

Figure 8.
Diffraction grating [71].

are projected at infinity due to the parallel collimated light [65–69]. The waveguide substrate is transparent which projects the virtual image without blocking the views from the real world.

SRG is an optical element with a periodic fluctuation structure at the wavelength scale on the surface of the material which can spatially modulate the light. VHG is the light–dark interference fringe formed by holographic technology exposure inside the material. They both cause a periodic change of the refractive index in the material. This periodic change is generally on the scale of the wavelength of visible light (450-700 nm), to allow effective manipulation of light for normal display [70, 71]. A single wavelength can be divided into several diffraction orders by the diffractive grating, and each diffraction order will continue to propagate in a different direction, as shown in **Figure 8**. The diffraction angle of each diffraction order is determined by the incident

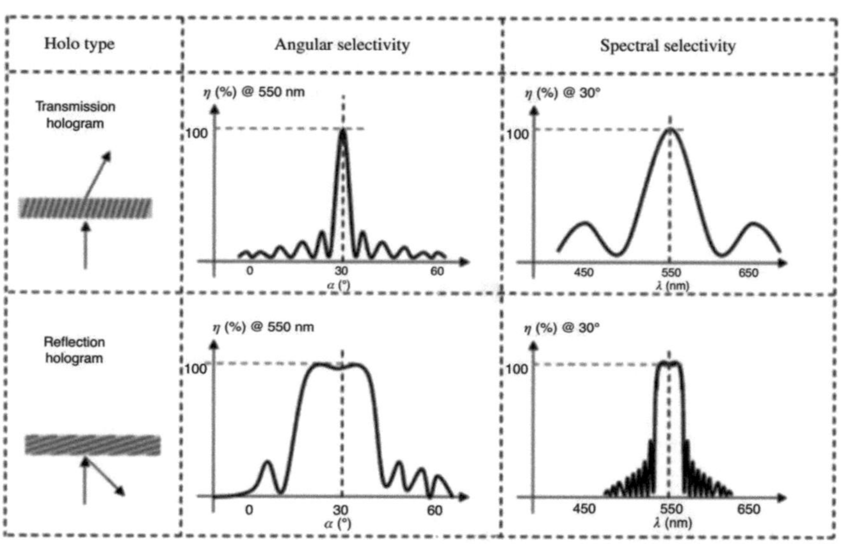

Figure 9.
Angular selectivity and spectral selectivity of transmission and reflection VHG [74].

angle of the light and the period of the grating. By optimizing associated parameters of the grating, such as the refractive index of the material, grating shape, thickness, duty cycle, etc., the diffraction efficiency of a certain diffraction order can be improved to the maximum, so that most of the light will keep propagating along this direction after diffraction [72, 73]. The diffraction modes include reflection diffraction and transmission diffraction, as shown in **Figure 9**. For the same grating period, the diffraction angles of different wavelengths are also different. Usually, there are three wavelengths applied for imaging, including red, green, and blue (RGB) elements.

However, the diffraction efficiency of the same color fluctuates depending on the incident angle, resulting in a different proportion of red, green, and blue light across the entire field of view (FOV), which is called the rainbow effect. This is caused by the inherent physical characteristics of diffraction gratings, and the color uniformity problem can only be optimized but not completely eliminated by design. Besides, the optical efficiency of the diffractive elements is very low, with nearly 85% of the projected light blocked, so only a small part can reach the human eyes. As a result, usually, an extra shading lens is needed for the observer to see the virtual images clearly and the contrast of the whole system is also reduced.

2.2.2 Geometrical waveguide

The reflective coating film is used as the in/out-coupler for the light propagating. The light is totally internally reflected by the in-coupler and bounces within the waveguide. When the light hits the out-coupler, part of the light is reflected out of the waveguide to human eyes while another part transmits through it for further propagating to the next out-coupler, as shown in **Figure 10** [75]. The partially reflected mirror array (PRMA) expands the exit pupil and achieves uniform light for virtual images.

Since the light will be reflected out from each out-coupler which makes the light density arriving at each out-coupler distinct, each of the out-couplers in the array needs a different reflection and transmission ratio (R/T) to ensure that the amount of light coupled out within the whole range of the eye pupil is uniform. The wavelength will not affect the imaging quality, but the R/T ratio would result in fringes in light and dark. The PRMA replicates the pupil to increase the total exit pupil area, however, it reduces the amount of light at each exit pupil area. Thus, the efficiency of the geometrical waveguide is less than that of conventional optical systems. Compared with the diffractive waveguide, manufacturing of the geometrical waveguide is more complicated, as

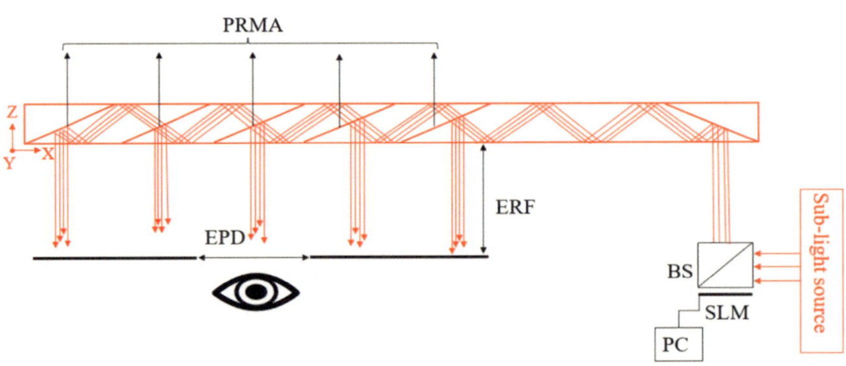

Figure 10.
The schematic of the geometrical waveguide [75].

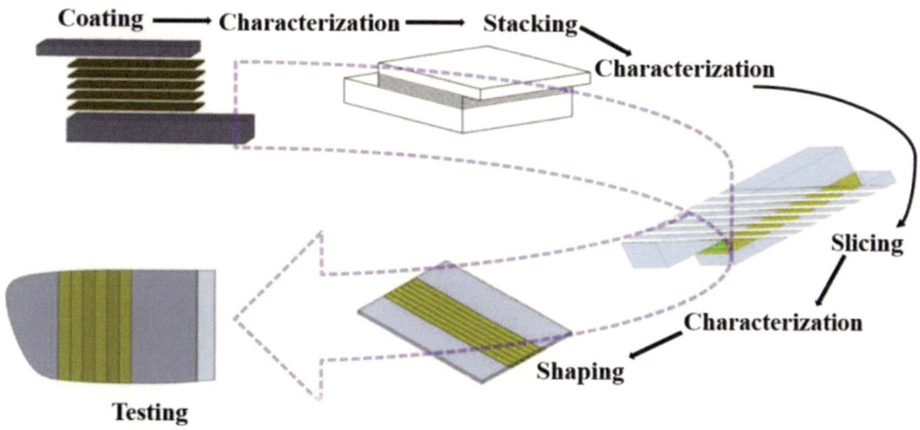

Figure 11.
The manufacturing process of the geometrical waveguide.

shown in **Figure 11**. It presents rigorous requirements to each step of the process chain for high manufacturing precision, usually at the micro and nanoscale.

Waveguide technologies have distinctive advantages in thickness and FOV, which makes them become a mainstream solution in AR display rapidly. **Table 1** is a comparison between diffractive waveguide and geometrical waveguide.

2.3 Light field technology

2.3.1 Pinlight technology

Pinlight display is a novel design to offer a wide FOV, supporting a compact form factor. Andrew et al. demonstrated the feasibility through software simulation and prototype display [76]. The defocused point light source is directly encoded. The spatial light modulation (SLM) between the pinlight and the eye modulates every pixel of the light, as shown in **Figure 12**. The encoded light is projected to human eyes to create virtual images by the pinlight plane and the SLM. There is an array of small hexagonal pits on it that capture total internal reflection light and create a precise spot

Waveguide Category	Diffractive waveguide		Geometrical waveguide
Companies	Microsoft, Magic Leap, Waveoptics, etc.	Digilens, Sony, etc.	Lumus, Lochn, Lingxi, etc.
In/Out-coupler	SRG	VHG	Reflective film
FOV	60°	50°	60°
Eye box	12 mm x 11 mm	12 mm x 10 mm	12 mm x 10 mm
Thickness	2.65 mm	< 2.5 mm	1.5 mm
Eye relief	18 mm	19 mm	16 mm
Manufacturing	Lithography technology, nanoimprint	Holographic interference	Coating, stacking, slicing, polishing, shaping

Table 1.
Comparison between diffractive waveguide and geometrical waveguide.

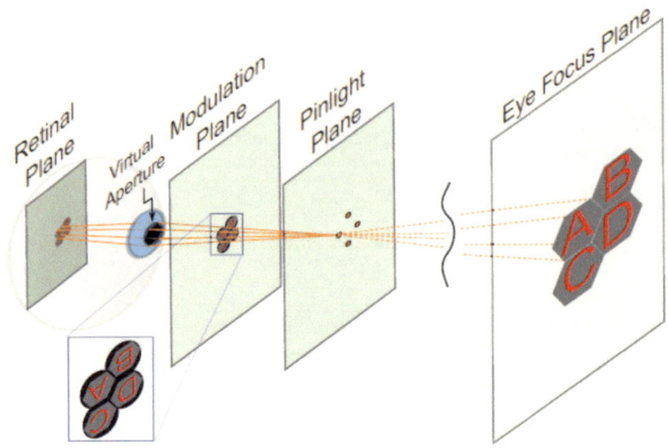

Figure 12.
Schematic of pinlight technology [76].

of light. People can observe the integrated virtual images by the coded series of light [76, 77]. However, the non-uniformity of the shape and intensity is still complex and difficult to be solved. The eye box and eye relief are also limited.

2.3.2 Pinhole technology

The pinhole technology has been applied in LetinAR. The vertically polarized light from the display is reflected by the beam splitter to human eyes. Polarized light can only enter the human eye after passing through the pinhole, and most of the light is blocked by the flat plate. Thus, the light efficiency is decreased due to the imaging principle. Pinhole technology can achieve smaller thickness and a high transparent ratio of the element [78, 79]. The eye box is related to the number of pinholes, and the virtual images can only be observed in a very limited eye relief range (**Figure 13**).

2.4 Comparison

Among the above technologies, FOV, eye relief, eye box, thickness, distortion, size, weight, and transmittance are important optical parameters to evaluate the

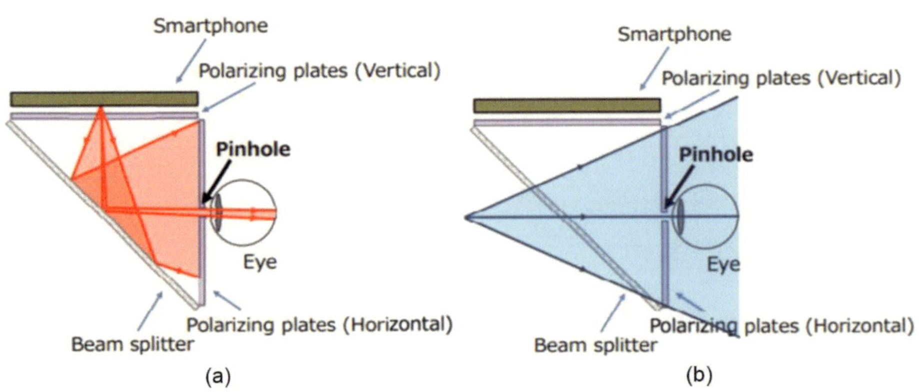

Figure 13.
Schematic of pinhole technology. (a) Red lines are from the projector. Most lines are locked by the polarizing plate and only the light passing through the pinhole can transfer to human eyes. (b) Blue lines are the light from real world [79].

Optical category	Conventional optics	Waveguide optics	Light field optics
FOV	Small	Large	Medium
Eye relief	Large	Large	Small
Eye box	Small	Large	Small
Thickness	Thick	Thin	Thin
Distortion	More	Less	Less
Size and weight	Heavy	Light	Light
Transmittance	Low	High	High
Manufacturing	Easy	Harder	Harder
Company	Meta, Epson, Google glass	Lumus, Lingxi, Microsoft	Creal

Table 2.
Comparison of the above imaging technologies.

imaging performance. There is a trade-off among these correlated parameters thus achieving a perfect system performance is impossible. Different applications may focus on different key optical indicators based on the specific expectation and may sacrifice some secondary optical parameters. The difficulty of manufacturing affects whether the system can be mass-produced or not. **Table 2** compares the above imaging technologies [80–82].

3. Projecting technology

With the development of augmented reality graphics, the optical performance of the projector also affects the imaging quality of the virtual images. There are some major parameters deciding the performance of the projector screen, such as the pixel, resolution, pixel per inch (PPI), and contrast. The pictures on the projector screen are made of millions of pixels. Each pixel is composed of three-color pixels, red, green, and blue (RGB). The variety of colorful pictures is projected by adjusting RGB. Resolution is the number of pixels. PPI is the number of pixels per inch on the screen. The higher the PPI, the smaller the pixel size, and the sharper it is. Screen contrast refers to the ratio between the brightness of black and white. The higher the contrast, the brighter and more colorful the images.

3.1 LCD

A liquid-crystal display (LCD) is a flat-panel display technology commonly applied in televisions and computers, etc. The first mass-produced LCD panel technology is twisted nematic (TN). As shown in **Figure 14**, when there is no electric field on the liquid crystal modules, the molecules in the LCD cell twist by 90 degrees. When the light from the environment and backlight passes through the first polarizer, the light is polarized and distorted by the liquid crystal molecular layer. When it reaches the second polarizer, it is blocked. Thus, the viewer observes the black screen. When there is an electric field applied to the liquid crystal molecules, they unravel. When polarized light reaches the liquid crystal molecular layer, the light transmits

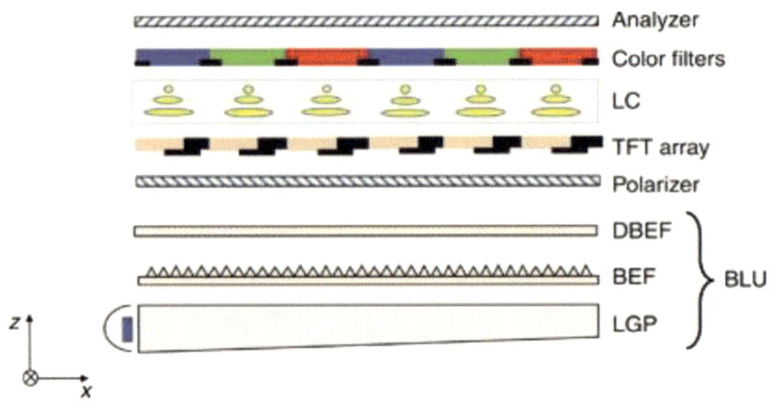

Figure 14.
Working principle of LCD [83].

directly without distortion. When it reaches the second polarizer, it can pass through, and the observed screen is bright. The electric field rather than the current makes the technology consume less power. By controlling the voltage, the deflection angle of the liquid crystal layer can be adjusted, and the brightness of each sub-pixel of RGB can be controlled respectively. By changing the brightness ratio, all colors can be realized by mixing RGB in different proportions. Due to the liquid crystal molecules cannot close completely, LCD cannot show pure black. LCD technology has the great advantage of being light, thin and low power consumption [83–85]. However, its drawbacks include slow response time (especially at low temperatures), limited viewing angles, and backlighting requirements.

3.2 OLED

A single Organic light-emitting diode (OLED) forms a pixel on the screen thus there are millions of OLED dots working together for the display. As shown in **Figure 15**, there is an organic light-emitting layer between two electrodes. When the positive and negative electrons crush in the organic material, the light emits. Each pixel of an OLED is composed of three sub-pixels, RGB, which light up when power is on. The brightness of each sub-pixel can be controlled by adjusting the voltage [86–88]. The brightness of the three colors can be mixed in different proportions to show the desired color. Since OLED lights themselves, no backlight source is needed, and each pixel can be independently controlled, enabling pixel-level light control. Compared with the LCD, OLED can provide pure black, and achieve no light leakage and perfect contrast. Besides, OLED screens have very short response time when switching between colors, with almost no drag. OLED screens are much thinner than LCD screens and can be bent considerably. OLED screens consume little power by switching on and off each pixel independently without all pixels working together. However, the lifespan of OLED is generally shorter than LCD by aging and burning problems due to the self-lighting mechanism. Moreover, there is an obvious strobe in low brightness, resulting in visual fatigue by pulse width modulation (PWM) dimming [89, 90]. Commercial cases include Epson is using their own OLED for AR glasses and SONY's OLED is applied for Nreal glasses [91].

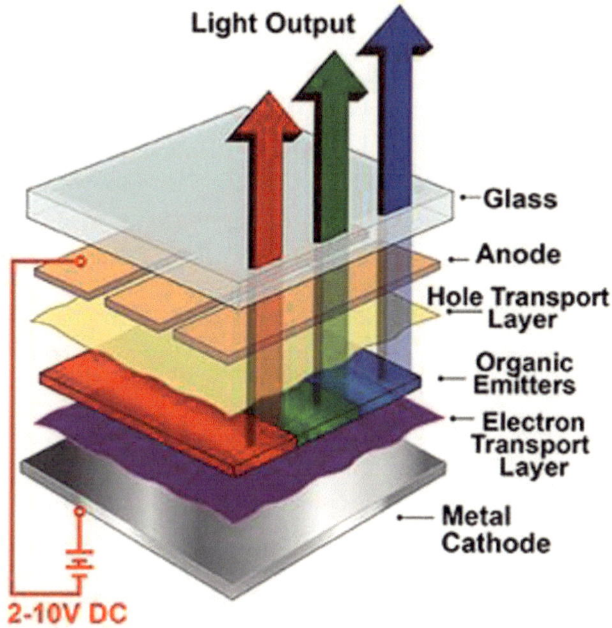

Figure 15.
Working principle of OLED [81].

3.3 QLED

Quantum-dot light-emitting diode (QLED) works based on the principle of quantum dots, which is to place quantum dots on a flat surface of a display and then use a control circuit to display the pictures, as shown in **Figure 16**. Quantum dots (QDS) have excellent optical properties, including continuously tunable peak positions of whole-spectrum luminescence, high color purity, and good stability, which are excellent luminescence and photoelectric materials. QLED display is built on these

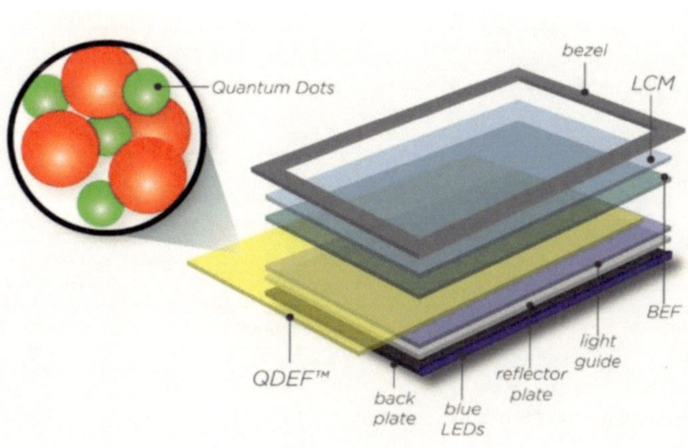

Figure 16.
Working principle of QLED [92].

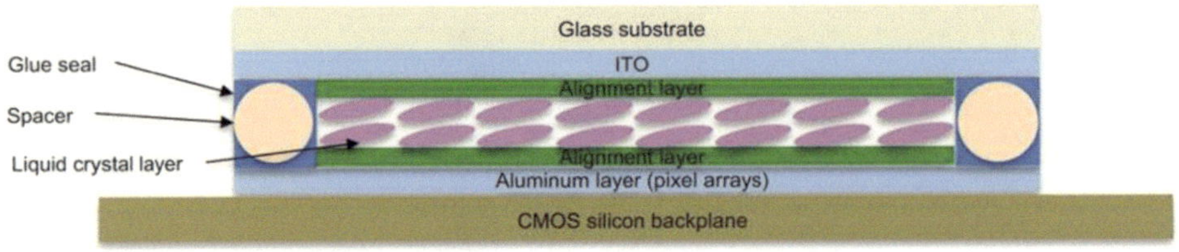

Figure 17.
Working principle of LCOS [95].

	LCD	OLED	QLED	LCOS
Substrate	**Glass**	**Glass/polymer**	**Glass**	**Glass**
Material of luminescent layer	LED/CCFL	Organic light emitting material	Quantum dot luminescent materials	LED/CCFL
Polaroid	Line polaroid*2	Circular polaroid	Line polaroid*2	Line polaroid*2
Backlight module	Yes	No	Yes	Yes
Color filter	Yes	No	Yes	Yes
Light panel	Yes	No	Yes	Yes
Emitting principle	Backlight (inactive)	Organic layer(active)	QD (inactive)	Backlight (inactive)
Contrast	>10,000:1	>1,000,000:1	>1,000,000:1	>1500:1
Angle of visibility	<150°, chromatic aberration	~180°, no aberration	~180°, no aberration	~180°, no aberration
Power consumption	High	Low	Low	High
Thickness	>1.2 mm	<1.5 mm	<1.5 mm	>1.2 mm
Service life	Long	Short	Short	Long
Cost	Low	High	High	Low
Mobile phone screen	Yes	Yes	Yes	No
Flat-panel screen	Yes	Yes	Yes	Yes
Working temperature	20°C ~ 60°C	−40°C ~ 80°C	40°C ~ 80°C	None
Impact on environment	Small	Big	Small	Small
TFT needed for each pixel	1	2	1	1
Toxic potentials	No	Yes	No	No
Spatial color uniformity	Great	Good	Great	Good
Light propagation	Transmission	Transmission	Transmission	Birefringence

Table 3.
Comparison among existing projector technologies.

special properties of QDS to achieve high performance and low cost. Compared with OLED, QLED can provide higher brightness and a wider spectrum, and the size is flexible with longer lifetime [92–94]. However, OLED has more advantages in no light

Key Merits	Ideal	Status quo
Brightness	$10^5 \sim 10^6$ nits	10^4 nits
Contrast Ratio (ANSI)	>300:1	<100:1
Refresh rate	>75 Hz	60Hz
Resolution	>60 pixel/DEG	30 pixel/DEG
Power consumption	<50 mW	100mW
Endurance	−55°C ~ 100°C	−10 °C ~ 60 °C
Form factor	Driver integrated, panel no larger than screen	Panel about 2*screen area
Emitting angle	Comply with exit pupil	Lambertian

Table 4.
Performance gap.

leakage, shorter response time, and wider visible view. Besides, OLEDs can provide a purer black display, better contrast, lighter and thinner, less power consuming, and performs better at night.

3.4 LCOS

Liquid Crystal on Silicon (LCOS) is a new reflective display technology combining LCD and CMOS integrated circuit. The birefringence characteristic of the liquid crystal molecule itself is applied for the light controlling. And the polarization of the incident light is modulated by the switch of the circuit to promote the rotation of the liquid crystal molecule. As shown in **Figure 17**, when the applied voltage of the LCD layer pixel is zero, the light does not enter the projection light path and there is no light output, that is, the pixel presents a dark state. When there is an applied voltage in the pixel, the bright is state, so the image is displayed on the screen. The voltage applied at both ends of the pixel will affect the optical performance of the liquid crystal molecule, and then determine the gray scale of the pixel. Its advantage is that it is mature and cheap, and its pixel density is relatively high, and the overall energy rate is also relatively high [95, 96]. Its disadvantages mainly lie in its relatively low contrast, especially at large incident angles, and it must be used in conjunction with PBS, which limits the miniaturization process of the overall optical system, and it cannot work at low temperature [95, 97, 98]. Mini Glass and Magic Leap use LCOS for the AR glasses.

3.5 Comparison

Table 3 shows the performance comparison among existing projector technologies. However, the state of the art cannot achieve ideal performance for better imaging quality in the optical see-through near-eye display. **Table 4** shows the performance gap between the ideal condition and the current craft. There is still a long way to go.

4. Conclusions

In recent decades, optical see-through near-eye display has undergone significant progress and facilitated various AR applications, especially for the industrial practice.

In this chapter, the display systems used by AR devices on the market, including various optical imaging systems and projecting systems, have been summarized. Among them, the difference in optical combinators is the key part to distinguish the AR display systems. The advantages and disadvantages of each optical technology are compared. Besides, the characteristics of the projectors are introduced and clarified its influence on the imaging performance and quality of the images. The gap to ideal performance is presented for the future development of relevant technologies.

Acknowledgements

This publication is partially supported by the Science Foundation Ireland under Grant Number 15/RP/B3208 and EIT Manufacturing 2022 KAVA 22125. For the purpose of open access, the author has applied a CC BY public copyright license to any Author Accepted Manuscript version arising from this submission. The acknowledgement also goes to the "111" Project by the State Administration of Foreign Experts Affairs and the Ministry of Education of China (No. B07014).

Author details

Jufan Zhang[1*], Yao Zhou[1] and Fengzhou Fang[1,2]

1 Centre of Micro/Nano Manufacturing Technology (MNMT-Dublin), School of Mechanical and Materials Engineering, University College Dublin, Dublin, Ireland

2 State Key Laboratory of Precision Measuring Technology and Instruments, Laboratory of Micro/Nano Manufacturing Technology (MNMT), Tianjin University, Tianjin, China

*Address all correspondence to: jufan.zhang@ucd.ie

References

[1] Berryman DR. Augmented reality: A review. Medical Reference Services Quarterly. 2012;**31**(2):212-218

[2] Azuma RT. A survey of augmented reality. Presence: Teleoperators & Virtual Environments. 1997;**6**(4):355-385

[3] Ong S, Yuan M, Nee A. Augmented reality applications in manufacturing: A survey. International Journal of Production Research. 2008;**46**(10): 2707-2742

[4] Zabels R, Osmanis K, Narels M, Gertners U, Ozols A, Rutenbergs K, et al. AR displays: Next-generation technologies to solve the vergence-accommodation conflict. Applied Science. 2019;**9**(15):3147

[5] Hua H, Javidi B. Augmented reality: Easy on the eyes. Optics & Photonics News. 2015;**26**(2):26

[6] Zhou Y, Zhang J, Fang F. Advances in the design of optical see-through displays. Advanced Optical Technologies. 2020;**9**:167-186

[7] Rolland JP. Wide-angle, off-axis, see-through head-mounted display. Optical Engineering. 2000;**39**(7):1760-1767

[8] Chi S, Yin Y, Yaoyuneyong G, Johnson E. Augmented reality: An overview and five directions for AR in education. Journal of Educational Technology Development & Exchange. 2011;**119**(4):119-140

[9] Yuen SC-Y, Yaoyuneyong G, Johnson E. Augmented reality: An overview and five directions for AR in education. Journal of Educational Technology Development and Exchange (JETDE). 2011;**4**(1):11

[10] Radu I. Augmented reality in education: A meta-review and cross-media analysis. Personal & Ubiquitous Computing. 2014;**18**(6):1533-1543

[11] Lee K. Augmented reality in education and training. TechTrends. 2012;**56**(2):13-21

[12] Moro C, Štromberga Z, Raikos A, Stirling A. The effectiveness of virtual and augmented reality in health sciences and medical anatomy. Anatomical Sciences Education. 2017;**10**(6):549-559

[13] Tang S-L, Kwoh C-K, Teo M-Y, Sing NW, Ling K-V. Augmented reality systems for medical applications. IEEE Engineering in Medicine and Biology Magazine. 1998;**17**(3):49-58

[14] Sielhorst T, Feuerstein M, Navab N. Advanced medical displays: A literature review of augmented reality. Journal of Display Technology. 2008;**4**(4):451-467

[15] Kamphuis C, Barsom E, Schijven M, Christoph N. Augmented reality in medical education? Perspectives on Medical Education. 2014;**3**(4):300-311

[16] Han DI, Jung T, Gibson A. Dublin AR: Implementing augmented reality in tourism. In: Information and Communication Technologies in Tourism. Cham: Springer; 2013, 2014. pp. 511-523

[17] Kysela JÍ, Torková P. Using augmented reality as a medium for teaching history and tourism. Procedia – Social and Behavioral Sciences. 2015;**174**:926-931

[18] Danielsson O, Holm M, Syberfeldt A. Augmented reality smart glasses in industrial assembly: Current status and future challenges. Journal of Industrial Information Integration. 2020;**20**:100175

[19] Livingston MA, Rosenblum LJ, Julier SJ, Brown D, Baillot Y, Swan II, et al. An augmented reality system for military operations in urban terrain. In: Proceedings of the Interservice / Industry Training, Simulation, & Education Conference (I/ITSEC '02), Orlando, FL, December 2-5, 2002. 2002

[20] Livingston MA, Rosenblum LJ, Brown DG, Schmidt GS, Julier SJ, Baillot Y, et al. Military applications of augmented reality. Handbook of Augmented Reality. 2011:671-706

[21] Mejías Borrero A, Andújar MJ. A pilot study of the effectiveness of augmented reality to enhance the use of remote labs in electrical engineering education. Journal of Science Education and Technology. 2012;**21**(5):540-557

[22] Piekarski W, Thomas B. ARQuake: The outdoor augmented reality gaming system. Communications of the ACM. 2002;**45**(1):36-38

[23] Stapleton C, Hughes C, Moshell M, Micikevicius P, Altman M. Applying mixed reality to entertainment. Computer. 2002;**35**(12):122-124

[24] Barakonyi I, Schmalstieg D. Augmented reality agents in the development pipeline of computer entertainment. In: International Conference on Entertainment Computing. Berlin, Heidelberg: Springer; 19 Sep 2005. pp. 345-356

[25] Ren D, Goldschwendt T, Chang Y, Höllerer T. Evaluating wide-field-of-view augmented reality with mixed reality simulation. In: 2016 IEEE Virtual Reality (VR). IEEE; 19 Mar 2016. pp. 93-102

[26] Lanman D, Luebke D. Near-eye light field displays. ACM Transactions on Graphics (TOG). 2013;**32**(6):1-10

[27] Kim S-B, Park J-H. Optical see-through Maxwellian near-to-eye display with an enlarged eyebox. Optics Letters. 2018;**43**(4):767-770

[28] Thibos L, Bradley A, Still D, Zhang X, Howarth P. Theory and measurement of ocular chromatic aberration. Vision Research. 1990;**30**(1):33-49

[29] Campbell FW, Gubisch RW. The effect of chromatic aberration on visual acuity. The Journal of Physiology. 1967;**192**(2):345-358

[30] Seidemann A, Schaeffel F. Effects of longitudinal chromatic aberration on accommodation and emmetropization. Vision Research. 2002;**42**(21):2409-2417

[31] Teo PC, Heeger DJ. Perceptual image distortion. In: Proceedings of 1st International Conference on Image Processing. Vol. 2. IEEE; 13 Nov 1994. pp. 982-986

[32] Schacter DL, Guerin SA, Jacques PLS. Memory distortion: An adaptive perspective. Trends in Cognitive Sciences. 2011;**15**(10):467-474

[33] Nicholls JI. The measurement of distortion: Theoretical considerations. The Journal of Prosthetic Dentistry. 1977;**37**(5):578-586

[34] Gu L, Cheng D, Wang Q, Hou Q, Wang Y. Design of a two-dimensional stray-light-free geometrical waveguide head-up display. Applied Optics. 2018;**57**(31):9246-9256

[35] Zhou Y, Zhang J, Fang F. Stray light analysis and design optimization of geometrical waveguide. Advanced Optical Technologies. 2021;**10**(1):71-79

[36] Stevens JC, Stevens SS. Brightness function: Effects of adaptation. JOSA. 1963;**53**(3):375-385

[37] Lotto RB, Purves D. The effects of color on brightness. Nature Neuroscience. 1999;**2**(11):1010-1014

[38] Arend LE, Spehar B. Lightness, brightness, and brightness contrast: 1. Illuminance variation. Perception & Psychophysics. 1993;**54**(4):446-456

[39] Starck J-L, Murtagh F, Candès EJ, Donoho DL. Gray and color image contrast enhancement by the curvelet transform. IEEE Transactions on Image Processing. 2003;**12**(6):706-717

[40] Hendrick RE, Mark HE. Basic physics of MR contrast agents and maximization of image contrast. Journal of Magnetic Resonance Imaging. 1993;**3**(1):137-148

[41] Ji T-L, Sundareshan MK, Roehrig H. Adaptive image contrast enhancement based on human visual properties. IEEE Transactions on Medical Imaging. 1994;**13**(4):573-586

[42] Zhou Y, Zhang J, Fang F. Vergence-accommodation conflict in optical see-through display: Review and prospect. Results in Optics. 1 Dec 2021;**5**:100160

[43] Zhou L, Chen CP, Wu Y, Zhang Z, Wang K, Yu B, et al. See-through near-eye displays enabling vision correction. Optics Express. 2017;**25**(3):2130-2142

[44] Lee S, Cho J, Lee B, Jo Y, Jang C, Kim D, et al. Foveated retinal optimization for see-through near-eye multi-layer displays. IEEE Access. 2017;**6**:2170-2180

[45] Park J-H, Kim S-B. Optical see-through holographic near-eye-display with eyebox steering and depth of field control. Optics Express. 2018;**26**(21):27076-27088

[46] Fontana U, Cutolo F, Cattari N, Ferrari V. Closed–loop calibration for optical see-through near eye display with infinity focus. In: 2018 IEEE International Symposium on Mixed and Augmented Reality Adjunct (ISMAR-Adjunct). IEEE; 16 Oct 2018. pp. 51-56

[47] Akşit K, Lopes W, Kim J, Shirley P, Luebke D. Near-eye varifocal augmented reality display using see-through screens. ACM Transactions on Graphics (TOG). 2017;**36**(6):1-13

[48] Zheng Z, Liu X, Li H, Xu L. Design and fabrication of an off-axis see-through head-mounted display with an x–y polynomial surface. Applied Optics. 2010;**49**(19):3661-3668

[49] Muensterer OJ, Lacher M, Zoeller C, Bronstein M, Kübler J. Google glass in pediatric surgery: An exploratory study. International Journal of Surgery. 2014;**12**(4):281-289

[50] Knight HM, Gajendragadkar PR, Bokhari A. Wearable technology: Using Google glass as a teaching tool. Case Reports. 2015;**2015**:bcr2014208768

[51] Deshpande S, Uplenchwar G, Chaudhari D. Google glass. International Journal of Scientific & Engineering Research. 2013;**4**(12):1-4

[52] Kress B. Optical design of next-gen augmented reality: Ahead of chairing the AR/VR/MR industry days at SPIE photonics west, Microsoft's Bernard Kress considers the optimal way of designing optics for augmented and mixed reality headsets. Electro Optics. 2019;**290**:12-17

[53] Wei L, Li Y, Jing J, Feng L, Zhou J. Design and fabrication of a compact off-axis see-through head-mounted display using a freeform surface. Optics Express. 2018;**26**(7):8550-8565

[54] Fang F, Zhang X, Weckenmann A, Zhang G, Evans C. Manufacturing and

measurement of freeform optics. CIRP Annals. 2013;**62**(2):823-846

[55] Cheng D, Wang Y, Xu C, Song W, Jin G. Design of an ultra-thin near-eye display with geometrical waveguide and freeform optics. Optics Express. 2014;**22**(17):20705-20719

[56] Wilson A, Hua H. Design and demonstration of a Vari-focal optical see-through head-mounted display using freeform Alvarez lenses. Optics Express. 2019;**27**(11):15627-15637

[57] Wang Y, Cheng D, Xu C. Freeform optics for virtual and augmented reality. In: International Optical Design Conference. Optica Publishing Group; 9 Jul 2017. p. JTu3A-1

[58] Fuge M, Yumer ME, Orbay G, Kara LB. Conceptual design and modification of freeform surfaces using dual shape representations in augmented reality environments. Computer-Aided Design. 2012;**44**(10):1020-1032

[59] Nikolov DK, Bauer A, Cheng F, Kato H, Vamivakas AN, Rolland JP. Metaform optics: Bridging nanophotonics and freeform optics. Science. Advances. 2021;**7**(18):eabe5112

[60] Huang H, Hua H. High-performance integral-imaging-based light field augmented reality display using freeform optics. Optics Express. 2018;**26**(13):17578-17590

[61] Zhou Y, Zhang J, Fang F. Design of a large field-of-view two-dimensional geometrical waveguide. Results in Optics. 2021;**5**:100147

[62] Lee Y-H, Yin K, Wu S-T. Reflective polarization volume gratings for high efficiency waveguide-coupling augmented reality displays. Optics Express. 2017;**25**(22):27008-27014

[63] Erdenebat M-U, Lim Y-T, Kwon K-C, Darkhanbaatar N, Kim N. Waveguide-type head-mounted display system for AR application. State of the art Virtual Reality and Augmented Reality Knowhow. 2018;**2018**:41

[64] Yoshida T, Tokuyama K, Takai Y, Tsukuda D, Kaneko T, Suzuki N, et al. A plastic holographic waveguide combiner for light-weight and highly-transparent augmented reality glasses. Journal of the Society for Information Display. 2018;**26**(5):280-286

[65] Waldern JD, Grant AJ, Popovich MM. DigiLens switchable Bragg grating waveguide optics for augmented reality applications. In: Digital Optics for Immersive Displays. Vol. 10676. SPIE; 21 May 2018. pp. 108-123

[66] Shin DH, Tibuleac S, Maldonado TA, Magnusson R. Thin-film optical filters with diffractive elements and waveguides. Optical Engineering. 1998;**1998**:37

[67] Roelkens G, Vermeulen D, Van Thourhout D, Baets R, Brision S, Lyan P, et al. High efficiency diffractive grating couplers for interfacing a single mode optical fiber with a nanophotonic silicon-on-insulator waveguide circuit. Applied Physics Letters. 2008;**92**(13):131101

[68] Chen CP, Mi L, Zhang W, Ye J, Li G. Waveguide-based near-eye display with dual-channel exit pupil expander. Displays. 2021;**67**:101998

[69] Akshaykore. Fundamentals of display technologies for Augmented and Virtual Reality. 2018. [Available from: https://hackernoon.com/fundamentals-of-display-technologies-for-augmented-and-virtual-reality-c88e4b9b0895

[70] Zhang Y, Fang F. Development of planar diffractive waveguides in optical

see-through head-mounted displays. Precision Engineering. 2019;**60**

[71] Palmer C, Loewen EG. Diffraction grating handbook. In 2004, Spectra-Physics, Including Richardson Gratings, Corion Filters, Hilger Crystals and Oriel Instruments, was acquired by Newport Corporation. 2005

[72] Palmer E, Hutley M, Franks A, Verrill J, Gale B. Diffraction gratings (manufacture). Reports on Progress in Physics. 1975;**38**(8):975

[73] Botten I, Craig M, McPhedran R, Adams J, Andrewartha J. The dielectric lamellar diffraction grating. Optica Acta: International Journal of Optics. 1981;**28**(3):413-428

[74] Kress BC, Meyrueis P. Applied Digital Optics: From Micro-Optics to Nanophotonics. John Wiley & Sons; 4 Nov 2009

[75] Zhou Y, Zhang J, Fang F. Design of a dual-focal geometrical waveguide near-eye see-through display. Optics & Laser Technology. 2022;**156**:108546

[76] Maimone A, Lanman D, Rathinavel K, Keller K, Luebke D, Fuchs H. Pinlight displays: Wide field of view augmented reality eyeglasses using defocused point light sources. In: ACM SIGGRAPH 2014 Emerging Technologies. 27 Jul 2014. p. 1

[77] Kramida G. Resolving the vergence-accommodation conflict in head-mounted displays. IEEE Transactions on Visualization and Computer Graphics. 2015;**22**(7):1912-1931

[78] Li Q, Deng H, Pang S, Jiang W, Wang Q. A reflective augmented reality integral imaging 3D display by using a mirror-based pinhole array. Applied Sciences. 2019;**9**(15):3124

[79] Katsumata Y, Yamada W, Manabe H. Optical see-through head-mounted display with deep depth of field using pinhole polarizing plates. In: The Adjunct Publication of the 32nd Annual ACM Symposium on User Interface Software and Technology. 14 Oct 2019. pp. 102-104

[80] Lee YH, Tan G, Yin K, Zhan T, Wu ST. Compact see-through near-eye display with depth adaption. Journal of the Society for Information Display. 2018;**26**(2):64-70

[81] Yang J, Twardowski P, Gérard P, Fontaine J. Design of a large field-of-view see-through near to eye display with two geometrical waveguides. Optics Letters. 2016;**41**(23):5426-5429

[82] Hu X, Baena FRY, Cutolo F. Alignment-free offline calibration of commercial optical see-through head-mounted displays with simplified procedures. IEEE Access. 2020;**8**:223661-223674

[83] Chen H-W, Lee J-H, Lin B-Y, Chen S, Wu S-T. Liquid crystal display and organic light-emitting diode display: Present status and future perspectives. Light: Science & Applications. 2018;**7**(3):17168

[84] Berreman D, Heffner W. New bistable cholesteric liquid-crystal display. Applied Physics Letters. 1980;**37**(1):109-111

[85] Hasebe H, Takatsu H, Iimura Y, Kobayashi S. Effect of polymer network made of liquid crystalline diacrylate on characteristics of liquid crystal display device. Japanese journal of Applied Physics. 1994;**33**(11R):6245

[86] Geffroy B, Le Roy P, Prat C. Organic light-emitting diode (OLED) technology: Materials, devices and display

technologies. Polymer International. 2006;**55**(6):572-582

[87] Gather MC, Köhnen A, Meerholz K. White organic light-emitting diodes. Advanced Materials. 2011;**23**(2):233-248

[88] Kay-khosa. OLED Display and It's Types. 2017. [Available from: https://steemit.com/steemstem/@kay-khosa/oled-display-and-it-s-types

[89] Tyan Y-S. Organic light-emitting-diode lighting overview. Journal of Photonics for Energy. 2011;**1**(1):011009

[90] Kalinowski J. Organic Light-Emitting Diodes: Principles, Characteristics & Processes. CRC Press; 3 Oct 2018

[91] Xu C. Nreal: Ready-to-wear mixed reality glasses. In: Spie avr21 Industry Talks II. Vol. 11764. SPIE; 29 Mar 2021. p. 1176409

[92] Moynihan T. What are quantum dots, and why do I want them in my TV. WIRED com. 2015

[93] Jang HJ, Lee JY, Kwak J, Lee D, Park J-H, Lee B, et al. Progress of display performances: AR, VR, QLED, OLED, and TFT. Journal of Information Display. 2019;**20**(1):1-8

[94] He L, Fei M, Chen J, Tian Y, Jiang Y, Huang Y, et al. Graphitic C3N4 quantum dots for next-generation QLED displays. Materials Today. 2019;**22**:76-84

[95] Zhang Z, You Z, Chu D. Fundamentals of phase-only liquid crystal on silicon (LCOS) devices. Light: Science & Applications. 2014;**3**:e213

[96] Lazarev G, Hermerschmidt A, Krüger S, Osten S. LCOS Spatial Light Modulators: Trends and Applications. Optical Imaging and Metrology: Advanced Technologies. 22 Aug 2012. p. 1

[97] Jülch V. Comparison of electricity storage options using levelized cost of storage (LCOS) method. Applied Energy. 2016;**183**:1594-1606

[98] Vettese D. Liquid crystal on silicon. Nature Photonics. 2010;**4**(11):752-754

Chapter 2

Smart Garment Design for an Augmented Reality Body Mapping Experience

Dominique Michaud, Geoffreyjen Edwards, Jocelyne Kiss, Jonathan Proulx-Guimond, Jonathan Caron-Roberge, Ernesto Morales, Mary Thaler and Cyril Schneider

Abstract

Based on previous work, we created wide handmade belts as support for a live augmented reality (AR) event. We describe how the garments were made to best meet the constraints of the AR design and how they were used as part of a fashion show and book promotion event. We determined the kind of features the AR software can detect and then created prototypes based on those characteristics. The belt design followed several steps: 1. Computer image preparation; 2. Thread preparation; 3. Thread dyeing; 4. Weaving; 5. Assembly; 6. Photographing the result and retouching the image. The lack of adjustability to deformation was a concern when dealing with clothing and fabric. We resolved this issue by stiffening the fabric. We discuss belts performances well during the live event; other textile techniques as substitutes (e.g. patchwork) and future research prospects.

Keywords: smart garments, augmented reality, body mapping, textile confection techniques, handweaving

1. Introduction

Plenum: The First Book of Deo [1] is a science-fiction novel written by Dr. Edwards. It presents the coming-of-age story of Vanu Francoeur, a novice in a monastic religious order who lives on a space station. In this distant future, society determines everything from birth—status, gender, and profession—and leaves little to individual choice. The novel also depicts numerous references to clothing and examples of visually changing garments, to reflect the wearer's emotions or state of mind. Our work here was to recreate four scenes from the book, to share how a smart garment can change one's body perception, that is, in a manner like what is sought in traditional body mapping, and as a complement to the use of the technology to promote the book itself.

We were inspired by the principle of body mapping [2], which is an art-therapy tool used to help patients express their feelings by drawing on a body-shaped cardboard that represents their own body. Drawing is a great substitute when words are hard to use, and extensive research has shown the utility of the approach [3–5]. However, this tool was designed for flat, 2D surfaces. Could something similar be achieved in 3D?

Augmented reality (AR) offers one possible solution. It mixes real and virtual elements to produce an immersive 3D experience. Drawings can be positioned as objects in the environment, and even their animation is possible. AR expands the realm of expressiveness available to patients. Even so, contrary to traditional body mapping, a camera is mandatory to get a view of the whole AR scene displayed inreal-time on a screen. This process has been described in previous studies by our team [6].

As visual AR is based on pattern recognition, it needs an anchor, a visual reference or "target." Considering the many references to clothing in the novel, we chose a garment to act as an anchor. We focused on the development of a wide belt as this is worn close to the body, can adapt to many shapes, covers the torso through a complete 360° rotation, thus is located on a relatively stable area of the body when walking, compared to the limbs or the head. To preserve the idea of having a garment that will evoke the novel and at the same time an impression of closeness and authenticity, we decided to create a handmade fabric. It was also easier to meet the technological constraints related to the use of AR if the belt was made specifically for this project.

Our main objective was to assess if our handmade fabric patterns could be recognized by Vuforia©, an AR software, and if we could generate stable images. There were many technical issues regarding the design of the belts; this is the focus of the paper. The main question was: To what extent is it possible to ensure efficient and robust pattern recognition using a handmade fabric as the target for the AR software? In the first part of this chapter, we present our methodology. We then describe how we proceeded to create the images that would be incorporated into the belts, create the fabric itself, and then constructed the belts. Next, we show how these innovations were used during a live streaming fashion show session. We conclude with a discussion of the technical limits of our process and future research perspectives.

2. Methodology

Vuforia is an AR software that detects target patterns, then allows augmented images based on this detection to be introduced. The images that appear in the AR mode, to be recognized and produced, must undergo a preparation phase with an online tool called Target Manager, available on the Vuforia website. Vuforia is then plugged into Unity3D, a virtual environment development software. We used Unity3D to display a 3D image or animation when the appropriate target was recognized, using trial and error to arrive at good results.

The first task was to determine the kind of features Vuforia can detect, the second was to create prototypes based on those characteristics. We tested first with paper templates printed from the Vuforia website to understand how the system works. We then created our own printed designs and tested these. After adjusting these to ensure good recognition capabilities, we experimented on handwoven fabrics. In the next section, we show step by step the process that led to the development of the first prototypes.

3. Garment design

As reported above, an earlier study [6] showed that AR can be achieved with a handmade fabric. The fabrication of the latter needs to comply with a set of rules to ensure that it can be used as an anchor for the animations. These rules are summarized below. Following this enumeration, we present the augmented images we developed.

3.1 Detection characteristics of AR software Vuforia

Vuforia is a software that can detect image features called targets; that is, it takes as input a camera image and looks for a specified target pattern. It then provides the possibility to overlay another image on top of those targets (or at a fixed offset), which produces the augmented image. Vuforia can be used with many platforms, including Unity.

Several image features were determined necessary to ensure good target detection:

- Asymmetry
- High contrast
- Clear edges
- Square image
- Many angled edges
- Black border
- No deformation.

Testing the different features on paper helped conclude that the best target is a black-and-white square image with many geometric straight-edge shapes on a rigid support.

The lack of adjustability to deformation from Vuforia is a concern when dealing with clothing and fabric. We resolved this issue by stiffening the fabric, first with cardboard, then with fusible interfacing. More details are provided in the following sections.

3.2 Prototypes

3.2.1 Weaving techniques and thread choices

The next step was to create and test fabric samples designed to maximize target recognition while at the same time serving the thematic focus of the project, that is, presenting scenes excerpted from Dr. Edwards's book. Many weaving techniques exist to create a fabric with colors that is asymmetric. These include jacquard, multi-layer, shadow weave, and crackle weave, the latter two of which require tweaking to avoid symmetry. Since the goal was to produce a figurative image, based on elements taken from the book, the jacquard weaving technique was deemed the most appropriate, as it offers the most versatility. It needs a specialty loom, however. There were two such machines to our knowledge in Quebec City. Unfortunately, the Covid-19 pandemic struck at that moment, rendering access to the jacquard looms difficult, so we sought another method.

One of us had a four-shaft weaving loom at home. This simple loom limits the kind of techniques one can achieve in a timely manner, but there are ways to produce a working fabric with the application of some creativity. It was thus decided to use the four-shaft loom to create samples. As for technique, both shadow weave and overshot were used to produce big and edgy patterns.

We also had to choose which kind of thread to use. Since we had to stiffen the fabric anyway, we decided to go with cotton, as it does not stretch, is easily available, and is cheap. Cotton of different counts was ordered, namely, 4/8 (coarse), 2/8 (medium), and 2/16 (fine), in black and white.

3.2.2 Weaving and finishing protocol

The thread was warped to obtain an alternation of black and white. Overshot threading was used with different setts ranging from 18 to 48 ends per inch with the same 12 dents per inch reed in all cases. The same number of picks and the same treadling were used for each of the two series of patterns: diagonals and semi-concentric diamonds. The first and last picks were woven into tabby. The initial test patterns were designed to combine the advantages of overshot and shadow weave, that is, to produce thick and contrasting outlines (**Figures 1** and **2**).

After light washing and drying, to allow the cotton fibers to take their final shape, the samples were fixed on a sturdy piece of cardboard to prevent the image captured by Vuforia from being distorted when handling the fabric (**Figure 3**). The choice to use the front or the back of the weave was guided by aesthetics. The weave was then photographed using a digital camera with natural light and no flash against a white background. The images were color-encoded in a JPEG format with a resolution of 180 pixels per inch. They were then cropped to keep only the area with the patterns (**Figures 4** and **7**). A static 3D object, a white cat, found in a free resource bank for Unity3D, was used to test the image augmentation (**Figure 6**). The detected points in the target patterns are shown as yellow crosses on the relevant images (**Figures 5** and **8**; **Figures 21–24**).

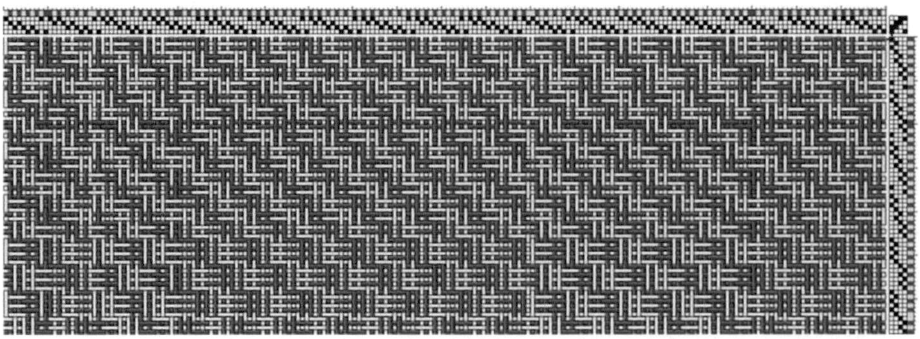

Figure 1.
Draft of diagonal pattern.

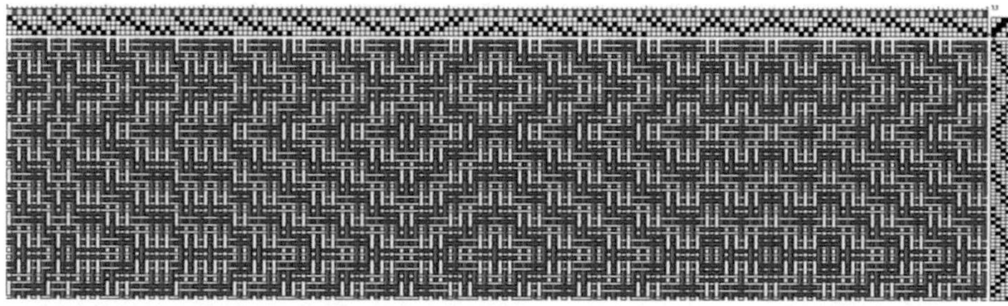

Figure 2.
Draft of semi-concentric diamond pattern.

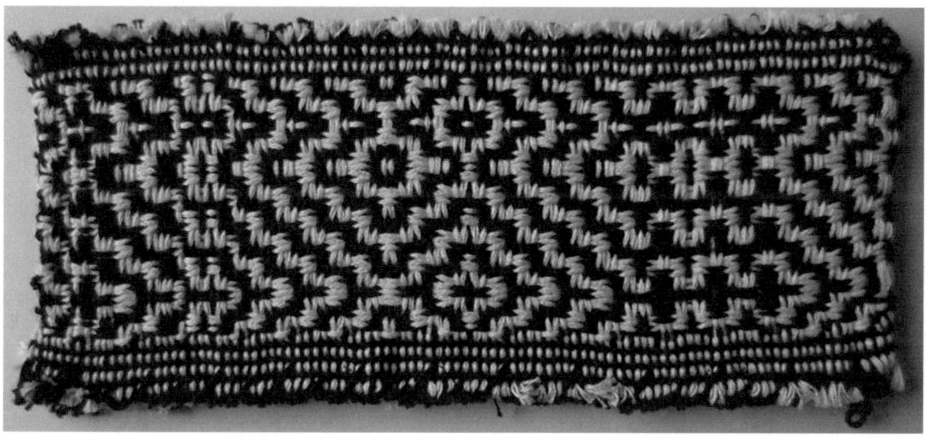

Figure 3.
Fabric of the diamond pattern after washing and fixing onto a cardboard backing, 4/8 thread, sett 18 ends per inch.

3.2.3 Target detection results

The Target Manager tool indicated that these weaves made solid, highly recognizable images with a multitude of well-distributed targets and a detection score of 5/5 for almost all weaves. Despite a rectangular aspect ratio (1:2), they performed well in augmentation tests with Vuforia on Unity3D. Lower detection scores were found when using medium and fine yarn at low densities, possibly because the aspect ratio

Figure 4.
Cropped image used with the target manager, diagonal pattern.

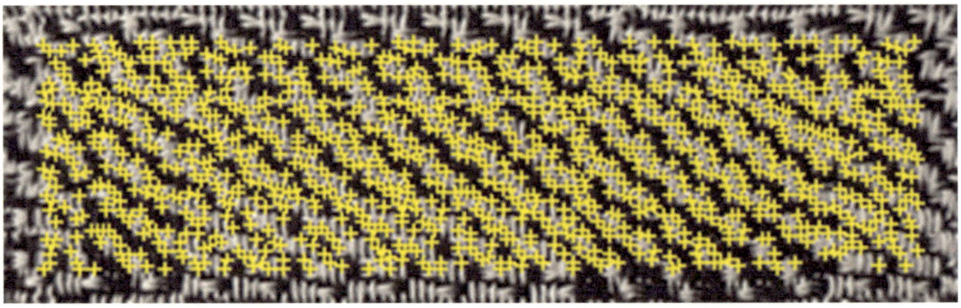

Figure 5.
Detected targets, diagonal pattern.

Figure 6.
Screenshot of augmented image with a digital white cat, diagonal pattern.

Figure 7.
Cropped image used with the target manager, diamond pattern, 4/8 thread, sett 18 ends per inch.

was higher. These results allowed us to decide in favor of the thicker 4/8 thread. The diagonal pattern detection score was 4 out of 5, which is very good (**Figure 5**). The diamond pattern, 4/8 thread, 18 thread count detection score was 5 out of 5, which is excellent (**Figure 8**).

Figure 8.
Detected targets, diamond pattern, 4/8 thread, sett 18 ends per inch.

4. AR system and smart Belt design

Dr. Edwards chose four scenes from his novel that served as inspiration for the design of the belts. These scenes were: 1. the surface of a moon whose resources are exploited by a space base built upon it (**Figure 9**); 2. a scorching sun with a space elevator (**Figure 10**); 3. A fractal drawn from the Mandelbrot set that acts as a dynamically changing body covering in the novel (**Figure 11**); 4. a space station within a nebula (**Figure 12**). Note that the lunar and solar surfaces were image subsets

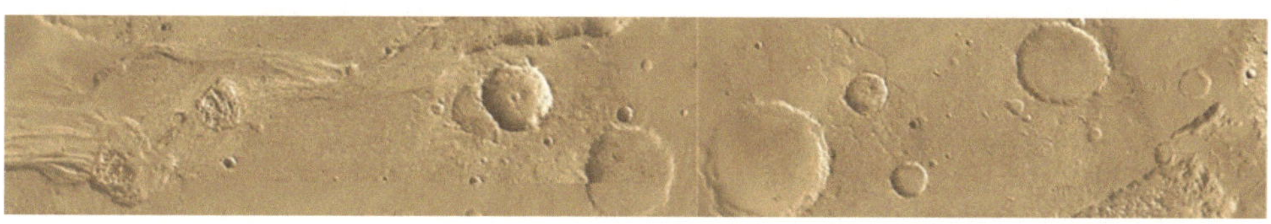

Figure 9.
Lunar belt original image.

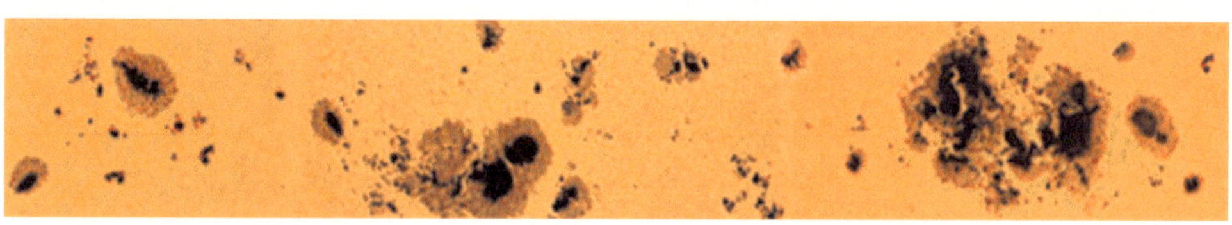

Figure 10.
Sun belt original image.

Figure 11.
Fractal belt original image.

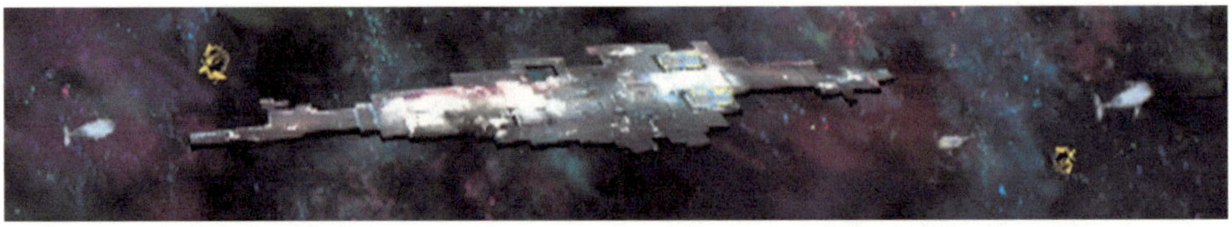

Figure 12.
Nebula belt original image; drawing by J. Proulx-Guimond.

from public domain NASA photographs obtained via the internet, the nebula scene resulted from an artist's rendition, and the fractal image was produced using a fractal algorithm running in Unity.

The belt design followed several steps: 1. computer image preparation; 2. thread preparation; 3. thread dyeing; 4. weaving; 5. assembly; 6. photographing the result and retouching the image. Note that we encountered difficulties with the detection algorithm applied to the nebula belt, due to a lack of contrast and the blurred outlines of the shapes. From a methodological point of view, this point caught our attention. This highlighted the importance of reinforcing the quality of the image and the color balance before its reproduction on the threads.

We used a double ikat-like dyeing technique [7] rather than the previous weaving techniques so that the whole set was more aesthetically consistent with the scenes in the book, even though the AR support does not need to have a conceptual link with the augmented image. We chose the ikat technique because it allows a great deal of precision in the dyeing process, as the warp and weft threads are painted separately. It produces an impression of both blurry and pixelated effects that lend themselves well to the aesthetic sought for this project and to the recognition constraints of AR. Ikat is a dyeing technique that consists of dyeing the threads before weaving. Everything must be well measured and calculated in advance to obtain a good reproduction of the patterns. However, when the preparation is liquid and therefore not very viscous, it tends to diffuse. A thickening agent, such as sodium alginate, agar-agar, and even starch, helps give a more precise contour.

The colors of the first three belts were chosen so that they stood out visually from each other. The blue color for the fractal belt was chosen when dyeing the threads to produce a chromatic contrast with the first two. The final belt was more colored to represent the diversity of colors that can be found when the nebula acts as the background.

4.1 Computer image preparation

At the very beginning of the AR process, the chosen image must be prepared on a computer. To do this, the image is cropped at a ratio of approximately 6:1 for width versus height. Used for a cylindrical shape, this ratio gives a rather square image, which is most effective for Vuforia. In practice, the height of the resulting belts was approximately 15 centimeters, which is the maximum that can be comfortably worn by a model just above the hips. Therefore, the length is about 90 centimeters, resulting in a diameter of 14.3 centimeters. This implies that our current belts can only be worn by slender models.

For the sun belt, sunspots were added where there were few elements to detect. The background was standardized, and the number of colors was reduced to four to accentuate the contrasts between light and dark areas. These modifications made it possible to obtain the most stable belt for an augmentation. The image was then adjusted to 12 pixels per inch in height and 12 pixels per inch in width, the same measurements as used in a test sample. The image was then printed at its actual size in two single-sided copies and one copy mirror-reversed over its length.

4.2 Thread preparation

One of the single-sided copies of the image was placed on a flat surface previously covered with a tarpaulin. It was then covered with a transparent plastic. This served as a reference for placing the colors later. The warp consisted of 100 threads, of which the first 12 and the last 12 were black and the others white. The black threads were not visible on the final piece and served as a guide for sewing the bias during finishing. They also served as a buffer zone when the dye diffused too much into the weft yarn. Following this, the warp was stretched across a flat surface over the printed image. The threads were spaced according to the planned sett, which was 12 ends per inch (**Figure 13**).

For the weft, a rigid cardboard support was used, the measurements of which corresponded to the width of the weave, considering stiffness and shrinking. Single-sided and mirror images were laid on either side of the cardboard, aligning them together. The weft was wound continuously around the support, taking care to respect the final picking (**Figure 14**). The weft ends needed to be clearly identified. The warp and weft yarns were soaked in 2% sodium carbonate solution and left to dry. The

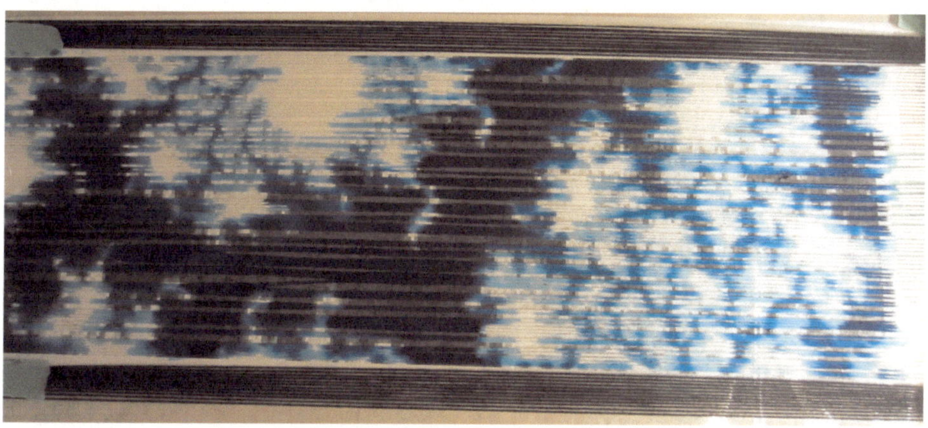

Figure 13.
Warp dyeing, threads stretched over a table, fractal belt.

Figure 14.
Weft dyeing, wrapped around the cardboard support, fractal belt.

Figure 15.
Loom weaving, fractal belt.

sodium carbonate served to increase the substrate pH, which allowed the chemical reaction with the dye.

4.3 Thread dyeing

The threads were dyed with reactive dyes as these are ideal for cellulose-based fibers. The dye was mixed in high concentration, between 4% and 8%, with urea water. The urea water was used to keep moisture in the fiber, long enough so that the chemical reaction can take place. To prevent the dye from spreading too much on the yarns, a thickener was also used. This was especially handy for adding small details. A dark dot was added on each end of the weft to indicate pick changes.

A tarpaulin, laid so that it did not touch the threads, was placed over the whole setup during the time the dye was soaking in. After that, the threads were removed from their holders, washed, rinsed, and dried.

4.4 Weaving

The warp was beamed onto the loom and threaded. A tabby section in white 4/8 thread of about two centimeters was woven at the beginning, and the end was overcast to prevent fraying; this was hidden when sewing the lining during finishing (**Figure 15**). Some unweaving was necessary to align the patterns.

4.5 Assembly

A horsehair interlining was used to give rigidity to the belt. This was heat-bonded onto the back, taking care to align the horsehair in the belt's height direction. The interfacing was inset by 1 centimeter at the top and bottom of the dyed area, that is, on the black border, along the width, and on the areas of white tabby at the weave's start and end. The excess fabric was folded over onto the back. A black polyester lining was sewn by hand onto the back. This lining ensured a finished look and was softer than the horsehair. Finally, a black polyester bias was also sewn in by hand (**Figure 16**).

Figure 16.
Belt assembly diagram.

A strip of black jersey about 30 centimeters long was sewn by hand to one end of the belt to serve as a fastening system. It could be fixed at the other end with pins. Located on the back of the user, this stretchy material facilitated breathing and adjustment to the body.

The bottom figure represents the hand-woven fabric ready for assembly. Above in beige is the horsehair interlining. Next comes the black lining, which covers everything and gives a finished look. At the very top, the black frame represents the bias that evens out the black border and completes the finish. The weaving is rectified using a square and a steam iron. One of the sides must be chosen to be the visible side. This is an aesthetic choice.

4.6 Photographing the result and retouching the image

Once the belts were constructed, it was necessary to photograph them and import the resulting image in the Target Manager software. For this, natural light was preferred. A white sheet, on which the belt was laid, was stretched out on a flat surface in front of a large window. The photo was taken vertically, avoiding shadows, and placing the belt as straight as possible. The image was imported into a photo editor to be cropped close to the black border. It was sometimes necessary to rework the border, such as for the lunar belt. In this case, the image was cropped flush with the dyed area and a new border was added (**Figures 17–20**). The retouched image was imported into the Target Manager website using the cylindrical model option. All images of the belts were placed in the same package that was imported into Vuforia (**Figures 21–24**).

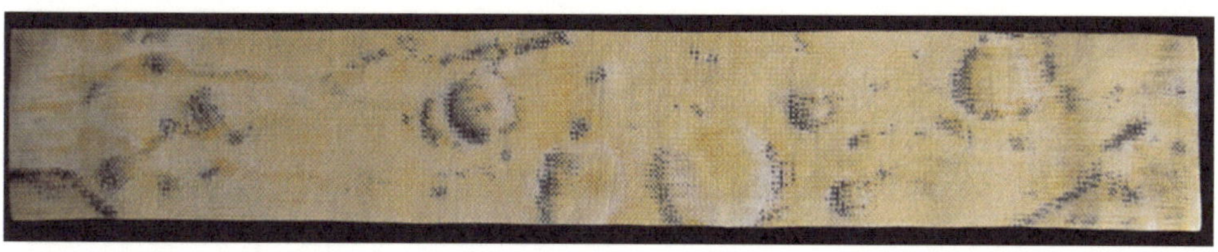

Figure 17.
Finished and retouched lunar belt image.

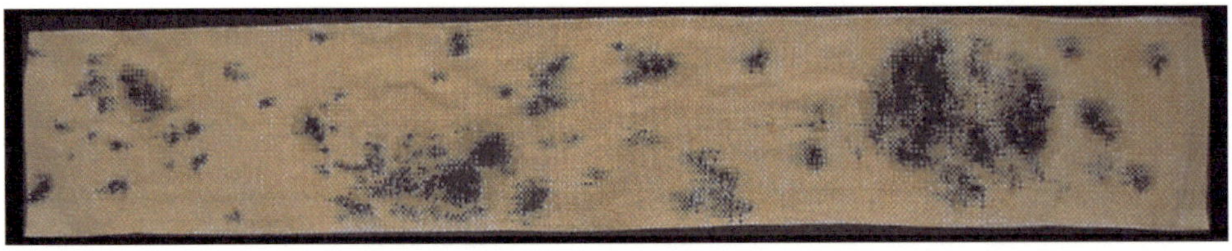

Figure 18.
Finished and retouched sun belt image.

Figure 19.
Finished and retouched fractal belt image.

Figure 20.
Finished and retouched nebula belt image.

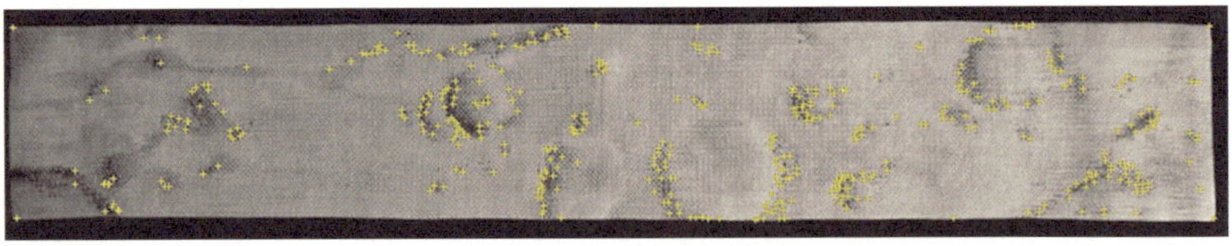

Figure 21.
Detected targets, sun belt.

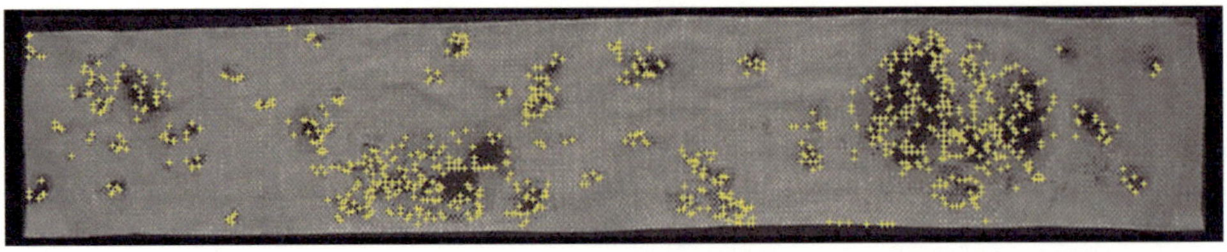

Figure 22.
Detected targets, fractal belt.

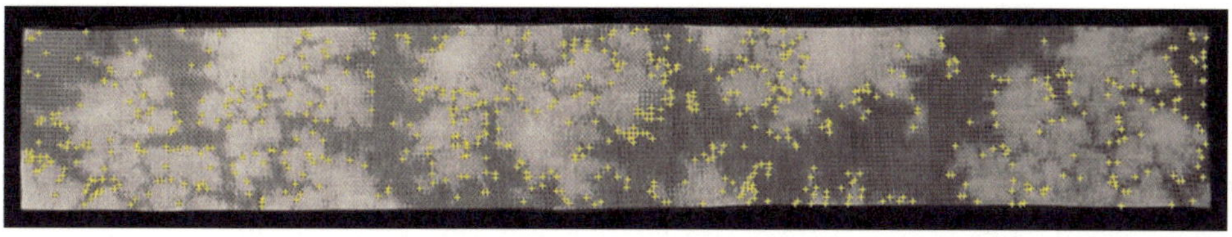

Figure 23.
Detected targets, moon belt.

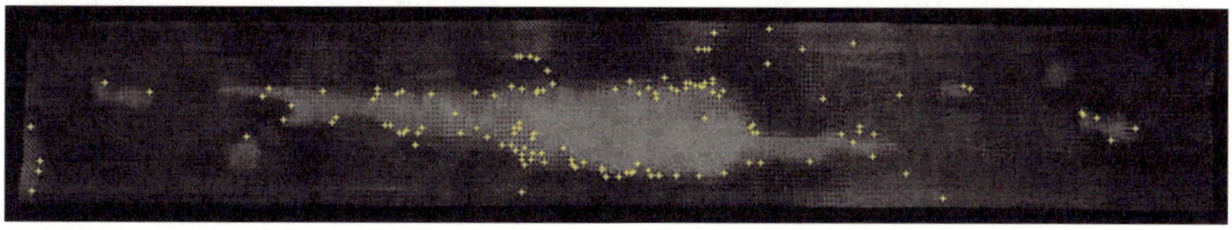

Figure 24.
Detected targets, nebula belt.

4.7 Software and digital development

Alongside the belt fabrication process, other team members carried out other parts of the project. Jonathan Proulx-Guimond was the animation designer. In particular, he drew the scenes from the novel and added several details that were not in the descriptions to make the scenes more complete. These details included the shape and texture of the ships, the look of the lunar docking station, the overall shape of the nebula space station, and the solar elevator. This creative work was carried out continuously, thanks to weekly meetings, which made it possible to adjust the project so that it formed a coherent whole. For example, for the moon belt, he was inspired by what D. Michaud, who did the weaving, had made beforehand to give a color theme and a texture to the moon, which was itself inspired by the images provided by Dr. Edwards. And conversely, the design of the nebula belt reflects the concept imagined by Dr. Edwards and Mr. Proulx-Guimond.

For the software component, Jonathan Caron-Roberge modeled the 3D scenes in Unity. He first created an application using Unity for users of body mapping integrating AR [2]. This application presents a character, male or female, standing in the common living area of a house. The character is seen in a third person view, without any characteristics, essentially a silhouette. In this version, the patient sees themself from the outside at first. The patient is then guided by a resource person mastering the application to add visual elements to their body map. The elements that can be added are mostly simple geometric shapes, such as cubes and spheres, but also flames and rain. It is possible to customize the color, size, and quantity of shapes as well as their position in relation to the body. Shapes can also move through space. Mr. Caron-Roberge then created another app for the show (see the next section). He incorporated Mr. Proulx-Guimond's animations and carried out the port of the project from Windows to iOS. This made it possible to use a smartphone to film the belts, transmit the information to a laptop computer to manage the image recognition and the augmentation, and send it back to the phone screen to view the rendering.

5. Live fashion show streaming

The design team carried out a fashion event showcasing the belts in real time, which was broadcast live on the internet, on April 23, 2022, from 2 p.m. to 4 p.m. (Quebec City time). This event, under the overall leadership of Dr. Edwards, was called Plénum À la Mode[1]. It brought together literature and fashion in an unprecedented hi-tech mash-up. During this event, our design team was almost all present.[2] We present the pictures of the belts taken that day (**Figures 25–28**). The resolution of the images is low as these are the images that were live streamed by Vuforia.

For the augmentation to be visible, the camera needed to be placed about 1 meter from the target, which can be difficult if the person wearing the belt is moving. This is a technical limitation of the recognition process. Therefore, the models were asked to stand still. Also, for stable augmentation rendition on the screen vis-à-vis the target, the latter needed to have a high detection score, which was not the case for all belts. Indeed, during preliminary experiments, it was determined that tracking the movement of a mannequin with a handheld camera, such as a smartphone, could lead to abrupt jumps in the final image. To counter this, we could have choreographed the movements. Unfortunately, this was not possible due to restrictions related to the Covid-19 pandemic. The different confinements and the limitations on the number of

Figure 25.
Model 1 wearing the moon belt with the moon station animation.

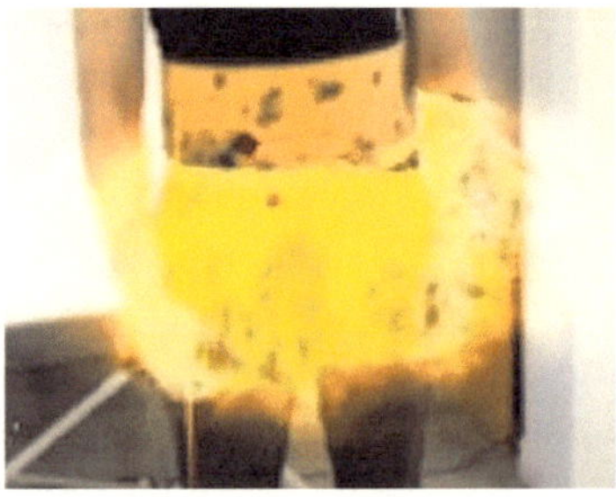

Figure 26.
Model 2 wearing the solar belt with the solar flare animation.

[1] Plénum À la mode, https://www.youtube.com/watch?v=l9VPQWNUtrI

[2] Geoffreyjen Edwards; Animation design: Jonathan Proulx-Guimond (absent); Original soundtrack: Jocelyne Kiss; Belt making: Dominique Michaud; Application development: Jonathan Caron-Roberge; Cameraman: Ernesto Morales; Models: Mary Thaler and Cyril Schneider.

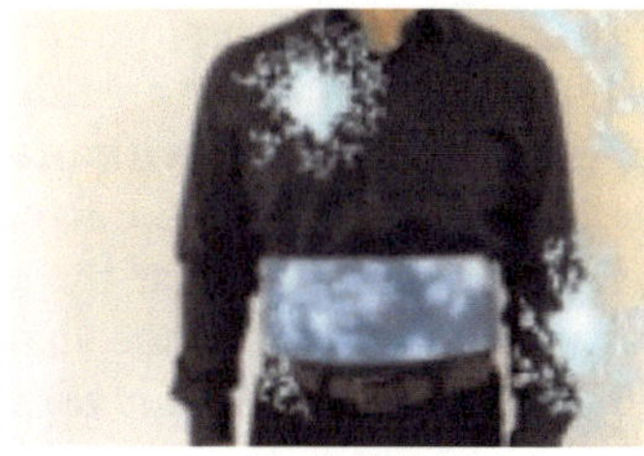

Figure 27.
Model 1 wearing the fractal belt with the fractal animation.

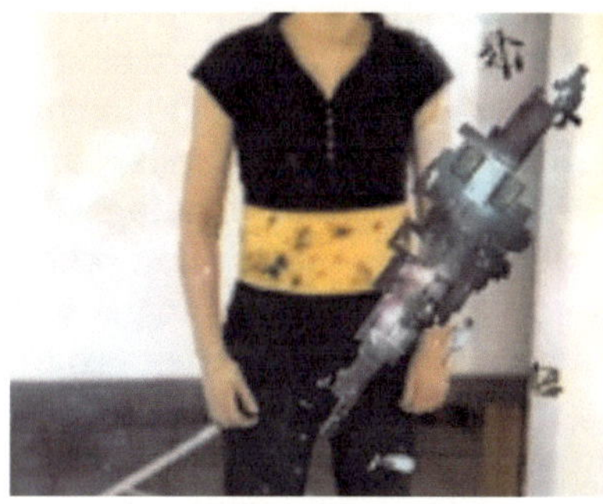

Figure 28.
Model 2 wearing the sun belt with the nebula animation and its station.

people who could gather in the same place meant that it was not possible to carry out rehearsals. As a result of these limitations, the models were instructed to stay put and maintain a relatively fixed posture.

6. Discussion and conclusion

The design and development of belts for body mapping produced with the assistance of a team and the organization of a live fashion show based on the technology demonstrated that AR is feasible using hand-woven fabric as the target pattern. Indeed, we achieved our objective, namely, to manufacture a garment linking virtual events with the bodies of the participants.

The sun belt performed the best. We believe this is due to the high contrast between paler and darker areas, the large number of target points all around the body, and a highly asymmetric pattern resulting from better pre-processing of the initial image (**Figure 22**). The black dye was the same for all the belts, but it was probably more concentrated on this one. The relatively uniform background orange color helped to make the black spots stand out better. The fractal belt was a close second, as far as target ratings for the Target Manager were concerned (**Figure 23**). The weft painting was quite precise and made for great contrasts despite a busy design. The moon belt did well, but contrasts were not as good, and some areas lacked targets (**Figure 21**). Granted, it was the first belt created and the whole production process was still under development.

The nebula belt, while the most pleasing visually, did not perform well with Vuforia (**Figure 24**). It was the last one made, and we did not have time to determine which elements were not working. The most possible reason is the lack of contrasts and the blurred edges. The design for this one was a deliberate stretch regarding target detection. Even though we experimented earlier with black-and-white samples, we wanted to try a more elaborate blending of the decorative elements. Regarding the cylindrical shape of the belts, it is worth noting that the augmentation was possible even though the models' torsos were not of the same shape or dimension as the one inputted to the Target Manager. The software seems to allow the 3D digital models some ease in terms of deformation (**Figures 25–28**).

It is also worth remarking that the patterns that can be created with a basic loom can be rich enough to be detected by a tool like Vuforia. The use of other weaving apparatus, such as a multi-frame loom or a jacquard loom, should also lead to useful and interesting results. Furthermore, the use of weaving as a base technique makes the virtual medium more concrete and the images more tangible.

Other textile-related techniques could be tested to achieve similar results. Patchwork is a sewing technique that consists of assembling small patches of fabric together. It is probably the best alternative to weaving as it can be done with little equipment. Embroidery involves sewing threads and other embellishments on fabric. We did try this on a few samples but not enough to get conclusive results. Knitting and crochet are unfortunately not an option, as they produce stretch fabric and hence induce deformations.

The next step could be to promote the fact that augmented reality on a textile support offers a promising avenue for art therapy via body mapping. This could take the form of a proof of concept with patients to measure the usability and usefulness of the setup. The integration of our first body mapping app, developed by Mr. Caron-Roberge, with the smart belt could also be undertaken. At the very least, this is a research project that we wish to explore in a future study.

Several uses are possible for such a technology. In addition to body mapping, consumer clothes might incorporate the technology to create interesting fashion statements. Some designers have started to explore this idea [8, 9]. It could also be possible to create tailor-made clothes, to create AR costumes for virtual universes of the mixed reality type to fit into gaming contexts or to share clothes between users of the same universe. We can also imagine using the technology for real-time performances, on a stage, for example, or for cosplay enthusiasts. In the field of entertainment, Disney Enterprises has recently developed an AR project that makes it possible to juxtapose an image of a digital costume on a person [10]. The associated patent describes how the image of the costume makes it possible to adjust to the static pose of a person, while respecting their personal morphology. There are other initiatives in this direction, namely, to superimpose a costume on a person. One blog post even defines this kind of application as "augmented costume" [11]. These AR applications have the advantage of allowing users to try on a costume, either for a one-off photo shoot or to visualize how it might fit their figure before making the physical garment. All these ideas have the disadvantage of requiring a stationary subject and a still pose. Our solution is innovative because a camera can move around a still-standing person and add animation to the other augmentations.

Modifications to the system as presented could also take different forms. For example, instead of only modeling the belt in the form of a cylinder, it would be interesting to represent the complete body of the user, as is the practice in traditional body-mapping techniques. A full body animated in real time would allow the augmentation to be adjusted to follow these movements. For this, the use of VuMark on the whole body (Vuforia's proprietary technology in the form of hexagonal targets), anchoring points on the garment to reveal virtual images, might be appropriate. This might also allow the use of other textile-manufacturing techniques, such as stretch fabrics and non-regular meshes.

Author details

Dominique Michaud*, Geoffreyjen Edwards, Jocelyne Kiss, Jonathan Proulx-Guimond, Jonathan Caron-Roberge, Ernesto Morales, Mary Thaler and Cyril Schneider
Laval University, Quebec, Canada

*Address all correspondence to: dominique.michaud.1@ulaval.ca

References

[1] Edwards G. Plenum: The First Book of Deo. Cosmos Cooperative; 2022

[2] Edwards G. Wearable technology and body mapping. In: Applying Body Mapping in Research: An Arts-Based Method. New York: Routledge; 2020. p. 12

[3] Gastaldo D, Rivas-Quarneti N, Magalhaes L. Body-map storytelling as a Health Research methodology: Blurred lines creating clear pictures. Forum Qualitative Sozialforschung / Forum: Qualitative Social Research. 2018;**19**. DOI: 10.17169/fqs-19.2.2858

[4] de Jager A, Tewson A, Ludlow B, Boydell K. Embodied ways of storying the self: A systematic review of body-mapping. Forum Qualitative Sozialforschung / Forum: Qualitative Social Research. 2016;**17**. DOI: 10.17169/fqs-17.2.2526

[5] Matos JAV, Silva KL, Garcia M-C. Body-map storytelling: Research experience report with theoretical contribution of Bourdieu. Esc Anna Nery. 2018;**22**. DOI: 10.1590/2177-9465-ean-2017-0407

[6] Edwards G, Caron-Roberge J, Michaud D, Guimond JP. Plenum a la mode - augmented reality fashions. In: Augmented Reality and its Application. London: IntechOpen; 2021. DOI: 10.5772/intechopen.99042

[7] Pebryani ND, Kleiss M, Sudharsana TIRC. Hidden calculation on patterning the warp and weft threads with double Ikat technique: Geringsing textiles. In: Proceedings of the 4th International Conference on Arts Language and Culture (ICALC 2019). Solo, Indonesia: Atlantis Press; 2020. DOI: 10.2991/assehr.k.200323.030

[8] Kim M, Cheeyong K. Augmented reality fashion apparel simulation using a magic Mirror. IJSH. 2015;**9**:169-178. DOI: 10.14257/ijsh.2015.9.2.16

[9] Häkkilä J, Colley A, Roinesalo P, Väyrynen J. Clothing integrated augmented reality markers. In: Proceedings of the 16th International Conference on Mobile and Ubiquitous Multimedia. Stuttgart Germany: ACM; 2017. pp. 113-121. DOI: 10.1145/3152832.3152850

[10] Guay M, Cimen G, Maurhofer C, Ryffel M, Sumner RW. US Patent for Automated costume augmentation using shape estimation. Patent (Patent # 10,885,708 issued January 5, 2021) - Justia Patents Search. 10885708. 2021

[11] Cosplay in Augmented Reality (AR). AR Critic 2019. https://arcritic.com/2846/cosplay-in-augmented-reality-ar/ [Accessed: July 8, 2022]

Chapter 3

Virtual Reality Utilization in Electrical Vehicle Development

Rares-Catalin Nacu and Daniel Fodorean

Abstract

This book chapter offers a new perspective on vehicle testing methods, by using driving simulators, which can be employed starting from early phases where components are designed, up to the real manufacturing stage. Based on these driving simulators, capable to thorough replicate real driving conditions, it can be described as a complex scenario, and the vehicle components can be dynamized with data coming from it. Therefore, with the use of a Virtual Reality (VR) environment, buildings, traffic signs, road types, and vehicles are imitated, and inspired by the real world or the imagination, but conclusive for the tested components. Moreover, for more realistic tests, a human driver can immerse into VR, controlling the virtual vehicle and leading to more reliable results. Many types of simulators are mentioned, with a focus on a specific type that is capable to test the vehicle's propulsion system and its driver assistance systems. In the end, a case study is exposed where different configurations, software, and hardware, are tested and several results are presented.

Keywords: virtual reality, driving simulator, electric vehicles, powertrain, human in the loop, hardware in the loop, co-simulation, EV modeling

1. Introduction

According to the latest report, published by the European Environment Agency [1], the case of premature deaths in the European Union (EU-28) for the year 2016, caused by suspended particulate matter ($PM_{2.5}$), nitrogen oxide (NO_2) and ozone (O_3), are estimated to 374,000, 68,000 and 14,000 of deaths, respectively. More nuanced, in terms of years of life lost (YLL), in the same causality order, the numbers are 3,848,000, 682,000, and 149,000 deaths. The most important sectors that contribute to the emission of air pollutants are transport, energy production and distribution, energy use in industry, industrial processes and product use, agriculture, households, institutions, and the commercial and waste industry. Relative to the year 2000, pollution mitigation is observed in all sectors, mostly in the transport sector, although both transported passenger and freight have been gradually increasing. Divided into two parts, non-road transports and road transports, both decreased the emissions of key pollutants (e.g. NO_x), but even so, this sector remains the most pollutant one.

From the road safety perspective, the number of victims due to road transport is amplified. Contributing factors in this increase are the infrastructure, the vehicles, the driving behavior, and the road exploitation. In the EU, in 2018, there was a threshold of 25,100 people who lost their lives while 135,000 were seriously injured. Compared to 2001, when the number of deaths was 54,000, there is a decrease of 55%. Considering the number of incidents in 2016 alone, of 1.35 million, supplementary measures need to be taken. Even so, the mindset, called "Vision Zero", is to move toward zero cases by 2050 [2].

In this context, restrictive anti-pollution measures (Euro 1 to Euro 6) affecting conventional vehicles are obvious, becoming increasingly stringent in the last two decades. At the same time, in some European countries, supporting financial aid are offered for purchasing electrified automobiles. This can only lead to an accelerated transition and development of alternative propulsion systems. Categorized by the powertrain configuration, there are hybrid-electric vehicles (HEV), plug-in hybrid electric vehicles (PHEV) electric vehicles (EV), and fuel cell vehicles (FCEV). Their manufacturing involves several challenges because it is a relatively new branch of industry and the alternative is expected to be, at least, at the same level of security and comfort as conventional ones. When it comes to HEV, PHEV, and EV, the challenges are the system configuration, energy storage capacity, vehicle mass, battery charging time, charging mode, choice of electric motor, etc. [3–7]. In case of FCEV, hydrogen storage is a significant impediment to their widespread adoption. Another consideration when selecting an unconventional car is the availability of charging stations, which are considerably less for these types of vehicles, especially for FCEVs, due to their low popularity, as well as, still, immature technology [8].

A significant feature of vehicles that would drastically decrease the number of casualties on roads would be the presence of advanced driver-assistance systems (ADAS). They convert conventional vehicles into intelligent vehicles (IV), using communication systems and sensors. Communication between vehicles (V2V), infrastructure (V2I), and pedestrians (V2P) and usage of radio detection and ranging (RADAR), light detection and ranging (LIDAR), and cameras, all these are the key elements to success, evolving from fully dependent vehicles to autonomous ones.

Considering the aforementioned, it is necessary to involve all technical and technological resources to accelerate the development of these types of vehicles, aiming for human life quality increase. A significant resource in this regard is the driving simulator. Through such devices and virtual reality (VR) technology, it is possible to mimic real driving conditions. Such systems are used to perform tests where the driver and vehicle behavior has to be observed. Frequently, the quality assessment of different infrastructure services is done in the same manner [9]. The researchers from [10] approached driving simulators from engineering, medicine, and psychological perspectives. From a medical and psychological point of view, the emphasis is on understanding the effects of alcohol, fatigue, drugs, ADAS, infrastructure, the weather over the driver's behavior, and also, the driver driving reactions. The engineering perspective emphasizes the development of the virtual world as real as possible so that the correspondence between the driving simulation and real driving response will remain almost unaffected. This means proper road designs, traffic signals and actors, and vehicle's dynamics. The goal is to have a reliable tool that will help the development of electrified vehicle power trains and their safety systems.

2. Purpose of the study

Generally, the lifecycle of a vehicle follows the V-Model, starting from the feasibility study where the concept is explored, and ending up with the operational phase, when the product is used. Each hardware and software component are verified, validated, and tested (VVT) in all phases including, definition, design, implementation, integration, qualification, and production, mitigating the risk of failure [11]. In case of vehicle components testing, what concerns the power train, most of the tests are made through different speed cycles, varying from one region to another [12–14]. In Europe, there are considered two classical road cycles for testing: the urban and extra-urban cycles, entitled "Urban Driving Cycle" (UDC) and "Extra-Urban Driving Cycle" (EUDC), respectively. In the U.S.A and Japan, there are "EPA Federal Procedure" (FTP-75) and "Japan Cycle 08" (JC08). Recently, a new group of speed profiles was adopted in Europe, Japan, India, and South Korea, entitled "Worldwide harmonized Light vehicles Test Cycles" (WLTC), classified by classes where each of them is divided into speed phases. Through the power-to-mass ratio (PMR) principle, there are three vehicle classes: WLTC class 1, (PMR $\leq$ 22 kW/ton), WLTC class 2 (22 kW/ton $<$PMR $\leq$ 34 kW/ton), and WLTC class 3 (PMR $\geq$ 34 kW/ton). In what it concerns speed, there are low, medium, high, and extra-high-speed phases. Although the new driving cycles are more realistic, still, some drawbacks are present. The retired EUDC comprises several steady-state speeds, which do not represent the behavior of a real driver. The Japanese cycle contains the behavior of the real driver, but only in congested city traffic. The new WLTC uses real driving speeds, for all types of roads, with different PMR, but still, some conditions are not taken into consideration, like for example the weather conditions, road surface, traffic, pedestrians, and road profiles [15–17]. These lack lead to an idealized, unrealistic test, and to a lack of correspondence between testing and operation, which can lead to defects, hazardous situations, or sizing problems of vehicle components.

In this chapter, different, relevant, and decisive testing is approached, through driving simulator platforms, using the VR concept. The use of such an instrument in the process of testing EVs leads to a true picture of its behavior during operation. Additionally, production time and costs are substantially reduced. Specifically, different component configurations with different parameters can be tested in conditions like the real ones, starting from the designing phase, without the need to build a prototype vehicle. The such approach offers the possibility to interfere within a specific phase of design if needed. Both major components can be tested, starting from hardware as the powertrain, chassis, sensors, electrical system, and software, such as control strategies for motors, energy management strategies, and ADAS.

3. Driving simulators

The definition of a car simulator can vary from a simple computer model that simulates the dynamic behavior of an element within the vehicle to a structure with multiple degrees of freedom (DOF), with high fidelity, accurately simulating real-time behavior. Depicted in **Figure 1** is a graphical representation of the most common types of car simulators, which were differentiated by the degree of flexibility/physical portability and the level of realism. Based on the simulation mode, a driving simulator can be defined as a model running offline or online.

3.1 The offline mode

This mode is represented by pure simulations, running on a computer where EV are mathematically modeled on software platforms such as *Matlab*, *Simulink*, *Amesim*

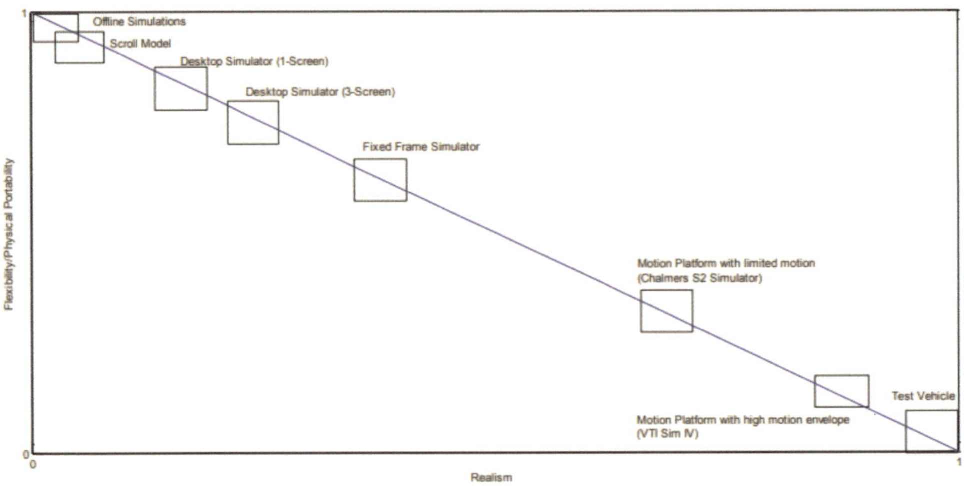

Figure 1.
Driving simulator hierarchy [18].

etc., with different input parameters for each component. For example, power train testing can be made using profiles for speed, slope angle, and wind. The vehicle model varies from one quarter, if a suspension unit represents the interest, to the entire vehicle. One of the limitations of offline simulation is that it does not have an interface or graphical representation to visualize the simulations while they are being performed.

With respect to **Figure 1**, from the same category, scroll models are another type of vehicle simulator that may or may not run in real time. Such a model allows the rapid change of input values through variables that can be associated with the pedals and/or the steering wheel, ensuring greater control during the simulation (**Figure 2**) [18].

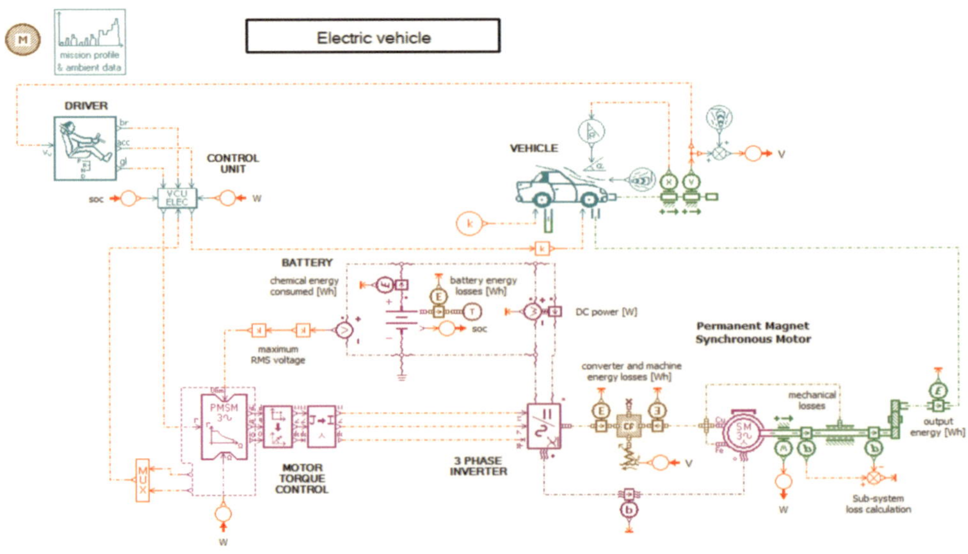

Figure 2.
Offline simulator model developed in Amesim.

3.2 The online mode

This mode refers to synchronization of a vehicle model with a real-time graphical interface, immersing the driver into a virtual reality environment. It contains at least one hardware unit, which is treated as the input for the model, manipulated by the driver, thus adding a man in the loop (MiL). Depending on the degree of complexity, the online mode covers a wide range of simulators, starting from a desktop-level simulator, to a motion simulator with multiple degrees of freedom [18].

One example of the online driving simulator was developed by the University of Jyväskylä, classified as a fixed frame simulator, which was used for a comparison study of the driver's behavior while driving a real car versus a simulator. It comprises a platform with two doors and a dashboard. The equipment for the Hardware in the Loop (HiL) analysis was a Logitech G25 steering wheel and pedals (**Figure 3**) [19].

Another simulator with the highest realistic level, called the "National Advanced Driving Simulator" (NADS), was developed for the US National Highway Traffic Safety Administration (NHTSA), by the University of Iowa. This is a simulator with a mobile platform with multiple movements that uses a cab with a vehicle body inside, and on the interior walls is projected an image with a 360° aperture. The body is located on four actuators that provide vibrations associated with rolling on different road surfaces, while the entire cab is located on a mobile platform. The latter can provide rotational movements on the x, y, z axes, lateral and longitudinal movements. Their role is to reflect the orientation and acceleration of the car inside the VR system, similar to real operation conditions (**Figure 4**).

Relevant scenarios for the traction chain can be simulated, based on real traffic conditions. In case of a traction motor, scenarios that implies different road types (e.g. urban, extra-urban, rural, motorway), with slopes, traffic, pedestrians, help in evaluating the dynamic response of the electric propulsion in all four quadrants. In cases where an overtaking is necessary, the motor is tested from the designing phase up to the real prototype, and its behavior is observed. Therefore, if needed, adjustments can be made from the early manufacturing process, leading to more suitable electrical propulsion systems.

Figure 3.
Driving simulator developed by the University of Jyväskylä [19].

Figure 4.
Outside (up) and inside (down) images from the NADS driving simulator [10, 20].

4. Driving simulator components

The focus of this sub-chapter is the development of an (efficient) EV, not the driving experience, and therefore the structure of the simulator will refer only to a fixed frame that is reliable enough to serve our purpose. The basic structure of this simulator consists of two main elements, indissoluble, which are represented by the software component and the hardware component. The software component allows the modeling and simulation of VR world, and the vehicle dynamics. Hardware support is also required to run these tools, represented by a computing system, which may or may not be real-time. Also, when necessary, the presence of input/output ports is mandatory, helping the user to interact with the simulator (**Figure 5**).

4.1 The software components

4.1.1 VR modeling software

One of the adequate software platforms which will do the job for the driving simulator was made by TASS and it is called *Prescan*. This software tool was specially designed for the development of ADAS systems and intelligent vehicles.

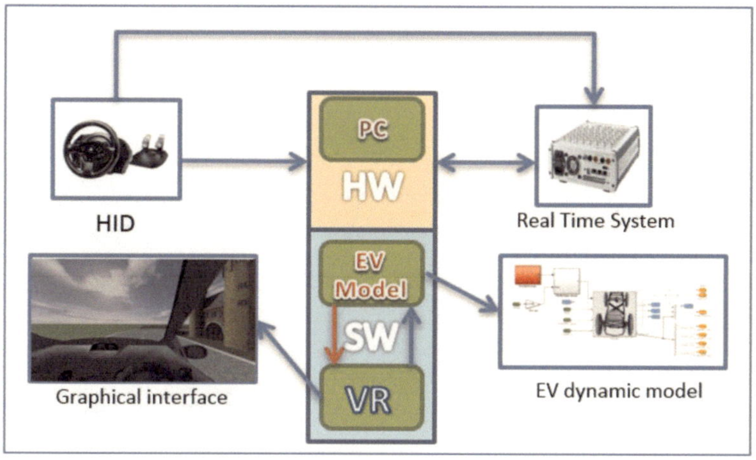

Figure 5.
Driving simulator structure with a fixed frame.

The usage sequence of the software depicted in **Figure 6** contains four major steps which must be followed prior to simulation evaluation. First, a relevant scenario for the interesting point is described, where the infrastructure and actors (vehicles) are modeled through Graphical User Interface (GUI), see **Figure 7**. The available library contains road segments, trees, traffic signs, buildings, and most important, the 'actors' (cars, motorcycles, and humans).

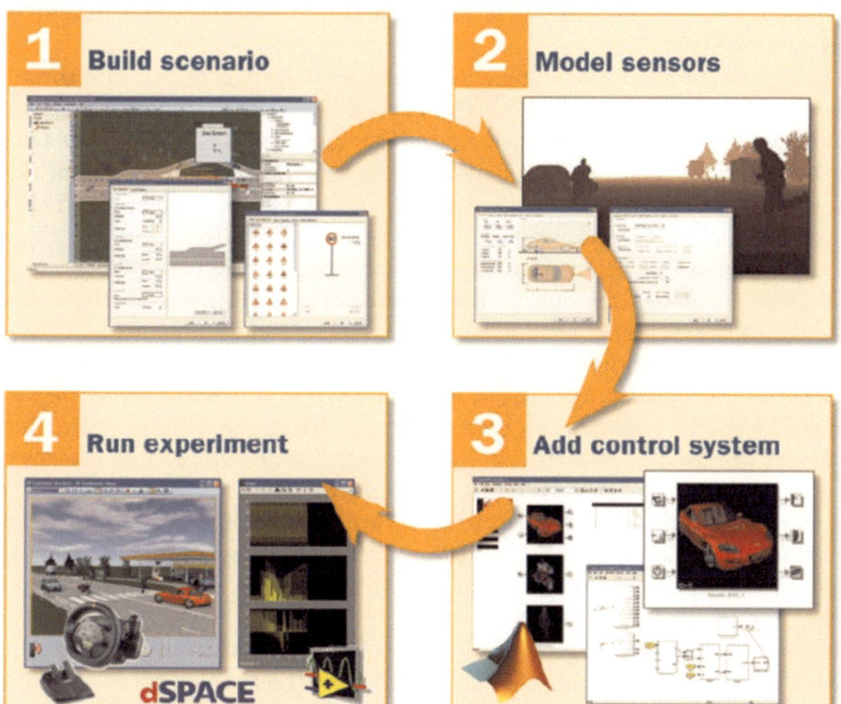

Figure 6.
Operation sequence of Prescan *VR software for real-time operation [13].*

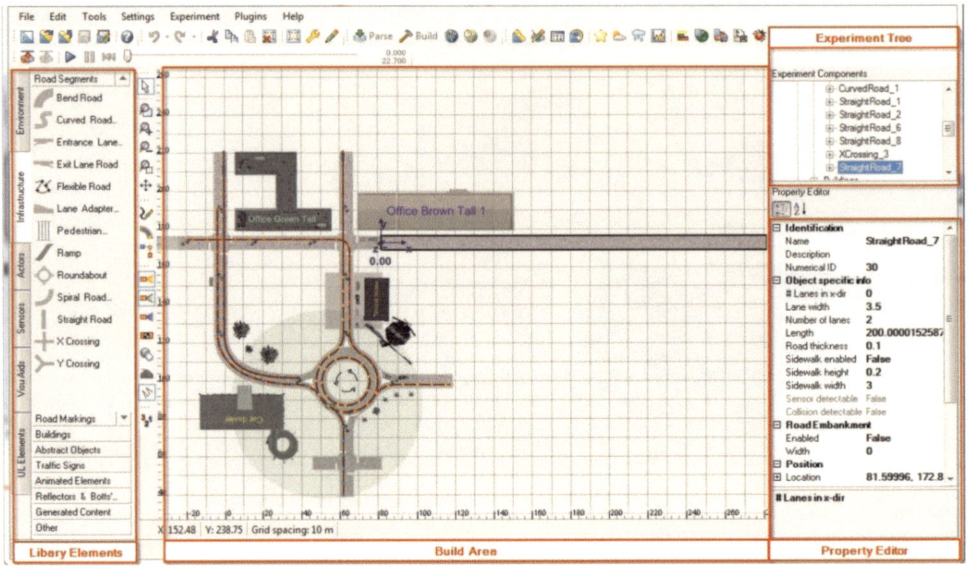

Figure 7.
GUI of Prescan *VR software interface [13].*

At this level of defining the infrastructure, which can be a fictitious one or replicated from reality, vehicle dynamics (inertias, body dimensions, tires, center of gravity, etc.) must be configured. At the same step, the actor's trajectories are defined, containing the paths and speed profiles. In car cases, the trajectory can be replaced with manual input, i.e. MiL, allowing the driving in the experiment through Human Interface Devices (HID).

The next phase is emphasized by the sensors modeling, grouped into three categories. Idealized sensors, which deal with V2I communication, using two principles, radio and infrared. There are detailed sensors such as cameras, LIDAR, RADAR, ultrasonic, and even Technology Independent Sensors (TIS). The last mentioned sensors, as the name says it, do not depend on any specific technology and are used to understand any active scanning sensor, offering the possibility to parametrize the beam, range, number of detected targets, target type, angles, energy loss, etc.

In the control stage of the operation sequence of the VR software, the mathematical behavior of the elements used within the experiments is introduced. If the previous two steps are defined in *Prescan* software, this stage of the study is employed in different software. Usually, a *Simulink* model will be sufficient to integrate the models; also, *Amesim* software can be considered. Within *Simulink*, specific input and output blocks are used to feed the VR experiment, but also to recover information about running conditions (for example, when the car is running on a hill, and the altitude is modified, the model needs to integrate such change within the mechanical model). This level of analysis is the most important since its complexity can affect the operation of the system and the accuracy of the results, but also the simulated driving experience and the reality perception.

In the last stage of the analysis, the experiment is running and the used can dive into virtual reality. This is again in *Prescan*, but it can be assisted by other real-time platforms for the real operation of equipment, such as electric propulsion, the energy transfer from the source etc. At this level, we are using a dSPACE platform, which is also programmed through *Simulink* to ease the interaction.

4.1.2 Vehicle dynamics modeling and computing platform

Several details will be given here about the third stage of the analysis. Before running the experiment, in "Engineering Workspace" represented by a different software platform as *Simulink* (in our case, but it can be in *Amesim* as well), the vehicle algorithms for sensors control and its dynamics are elaborated. The vehicle dynamic model is depicted in **Figure 8**, which contains the propulsion unit, shift logic, gearbox, and chassis. Based on throttle reference, motor dynamics, chassis dynamics, and the initial state of the vehicle, the next state is calculated, where state refers to position, velocity, and acceleration. Each state has two components, for translation, *XYZ*, while for rotation is *RPY*. The term *RPY* comes from *Roll*, *Pitch,* and *Yaw* angles, with rotational direction around *XYZ* axes.

4.2 Hardware components

A driving simulator which serves as a unit for EV power train research and development, is made from a frame that is support for a personal computer (PC) with one or more screens, a real-time (RT) platform, and a HID – see **Figure 9**.

The structure can be extended, when the real power chain is tested, with a real motor-generator configuration, a controller, a power source (battery), and an electronic load (to emulate the real electric power consumption). Moreover, in order to monitor the electric parameters, a measuring instruments cluster is interposed between the motor and controller, as between the generator and electronic load – see **Figure 10**.

There are three possible hardware configurations for our simulator, considering the EV dynamics. For the first two of them, the vehicle dynamics is deployed on computational platforms and the difference is whether it is or not in real-time, while the last one uses a real electric power chain (see **Figure 11** for the definition of the possible operating configurations).

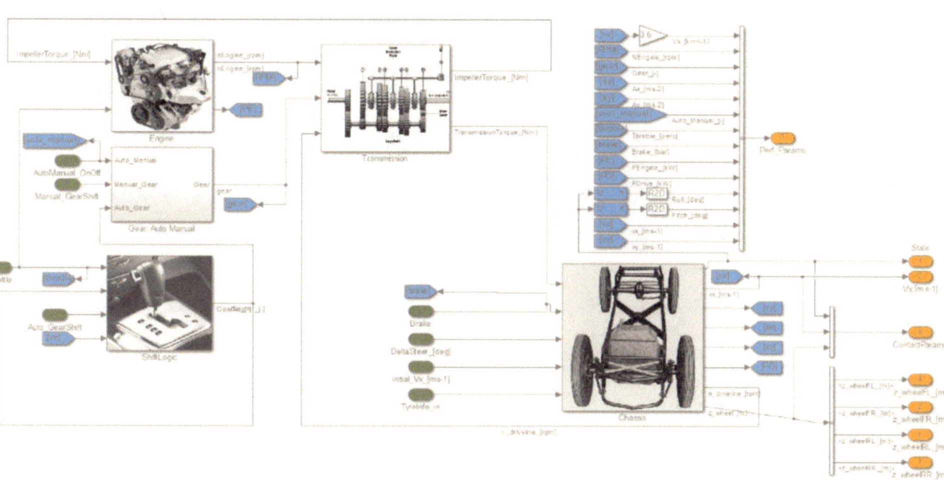

Figure 8.
Vehicle dynamic model developed in Simulink.

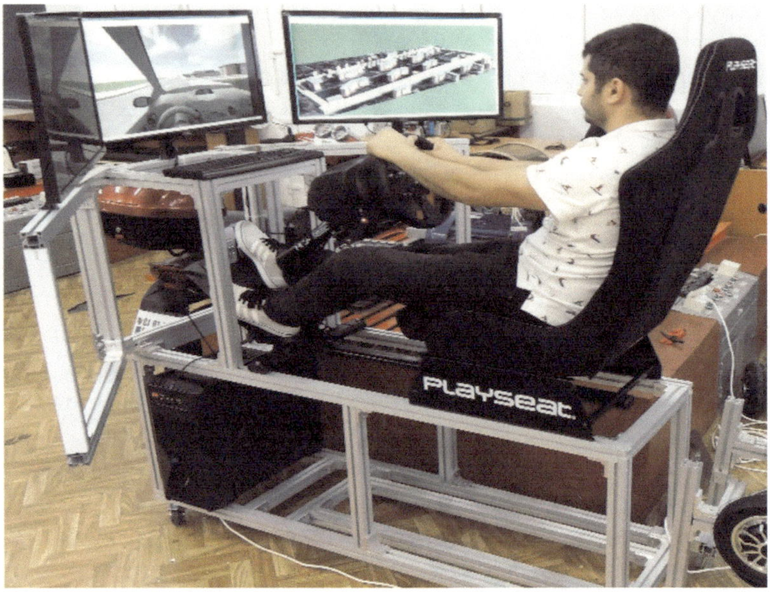

Figure 9.
The driving simulator developed at the technical University of Cluj-Napoca.

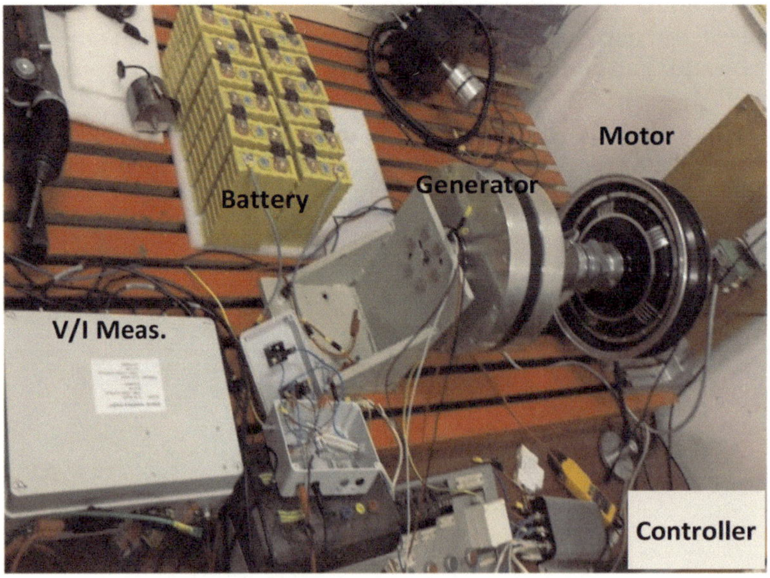

Figure 10.
The electric traction chain used to extend the developed driving simulator developed at the technical University of Cluj-Napoca.

4.2.1 HID-PC

The simplest configuration uses a PC and an HID, whereas the latest comprises a steering wheel and foot control pedals (gaming kit), connected via USB cable. The VR and vehicle dynamics computing software runs on PC, on the programmed scenario and driver input.

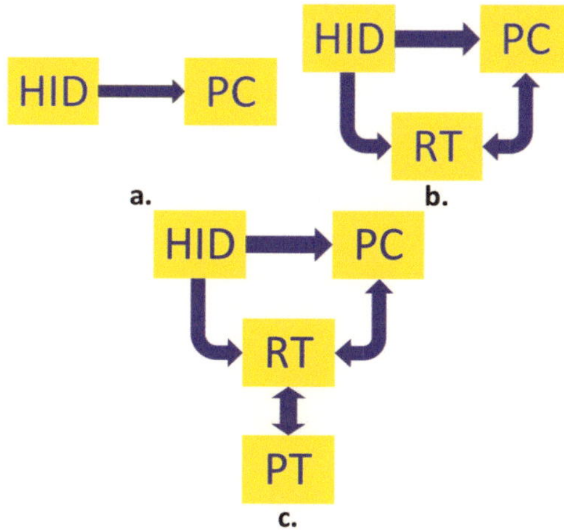

Figure 11.
Studied driving simulator hardware configurations.

4.2.2 HID-PC-RT

Additional to the previous configuration, there is an RT platform, for more complex vehicle models, where RT computing power is needed. The vehicle dynamic model of the power train is deployed on the RT platform, while the PC is used for chassis dynamics and VR. Here, the HID signals are divided, the path reference is sent toward the PC, while the speed reference is toward the RT platform. The resulting speed of the vehicle is sent to PC, based on reference, power train dynamics, and road angle.

4.2.3 HID-PC-RT-PT

The last configuration implies the presence of a real power train, consisting of two electric machines (one operating in the motor regime, while the second one in the generator regime), an electronic controller to control the electric motor, a power source (a battery) and an electric load (which recovers the electric power generated by the second electrical machine, based on a profile correlated with the running conditions which are imposing the total resistant force applied to the motor's shaft). The motor is controlled by the electronic controller, supplied from the battery, with the reference speed signal provided by HID. The energy produced by the generator could be injected back into the battery, supercapacitors, or into the grid. The motor speed is measured and sent to PC (within *Simulink*), and as a result, the vehicle moves into VR accordingly (within *Prescan*).

5. Driving simulator modeling

As mentioned above, several steps must be followed prior to the experiment launch, represented by the sequence:

5.1 Scenario designing

The world within the VR can be modeled by a fictitious scenario, elaborated on basis of testing interest points, or/and from reality, thus replication of the real world.

A good basis for the real world can be exported from *Open Street Map*, as.OSM file. This type o file contains data in form of points, connections, and street name properties (tags). Thus, the infrastructure is rebuilt in 2D, with no elevation option. Built from segments, if the altitude of each end of the segment is known, conversion into 3D is made possible. The altitude information can be obtained through *Google Maps*, in the *Terrain* section. Next, importing a base layer for road, which is exported from *Google Maps* as a Print Screen, helps with graphics improvement. Due to lack of information in exported. OSM file regarding the buildings and traffic signs, replication can be achieved through the software library.

The real and replicated maps used in the VR experiment are depicted in **Figure 12**.

The map must be populated with actors, represented by vehicles and humans, each of them having their own law of motion. After placing the vehicles, their dynamics are parametrized accordingly, starting from inertias, wheel radius, center of gravity, body dimensions, suspension stiffness and damping rate, ending with tire stiffness. Subsequently, vehicle and infrastructure sensors will be added and parametrized, depending on the sensor type. There are some generalities for all the sensors as position, orientation, and range, but also particularities, for example, a camera resolution and its focal length or the capture frequency for an ultrasonic sensor. In the end, for the actors which are not manipulated by HID, the trajectories are assigned to them. A sample of the employed parametrization of the aforementioned characteristics is shown in **Figure 13**.

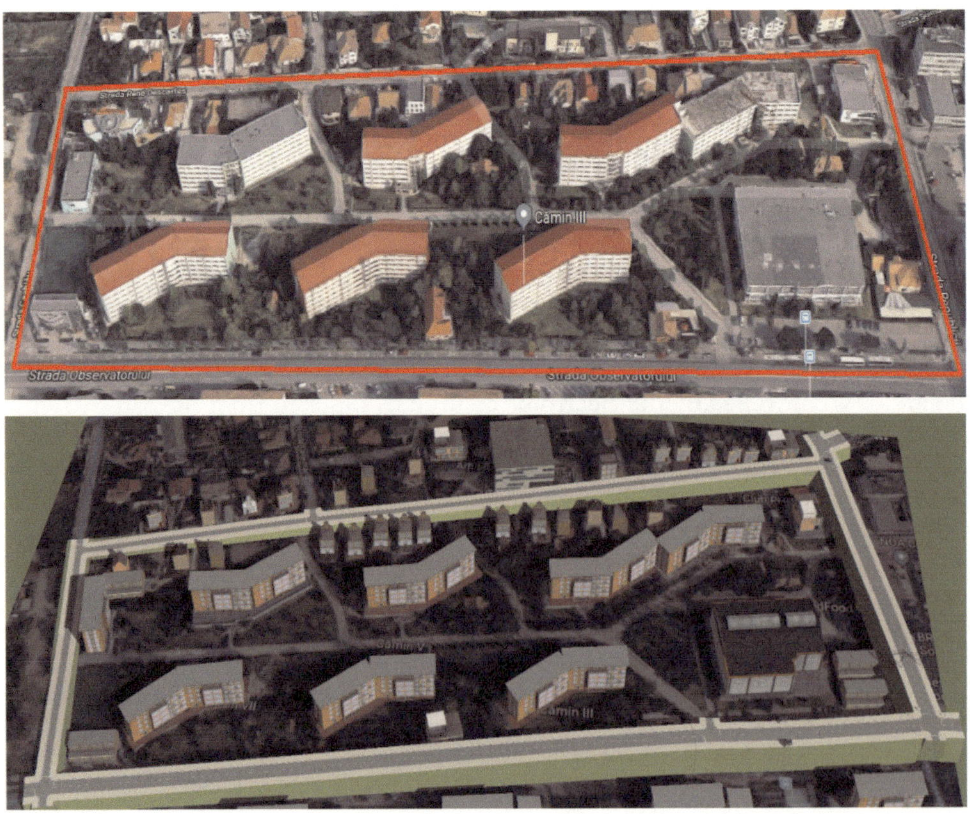

Figure 12.
Real Google *(up) and replicated (down) maps drawn in* Prescan *VR infrastructure.*

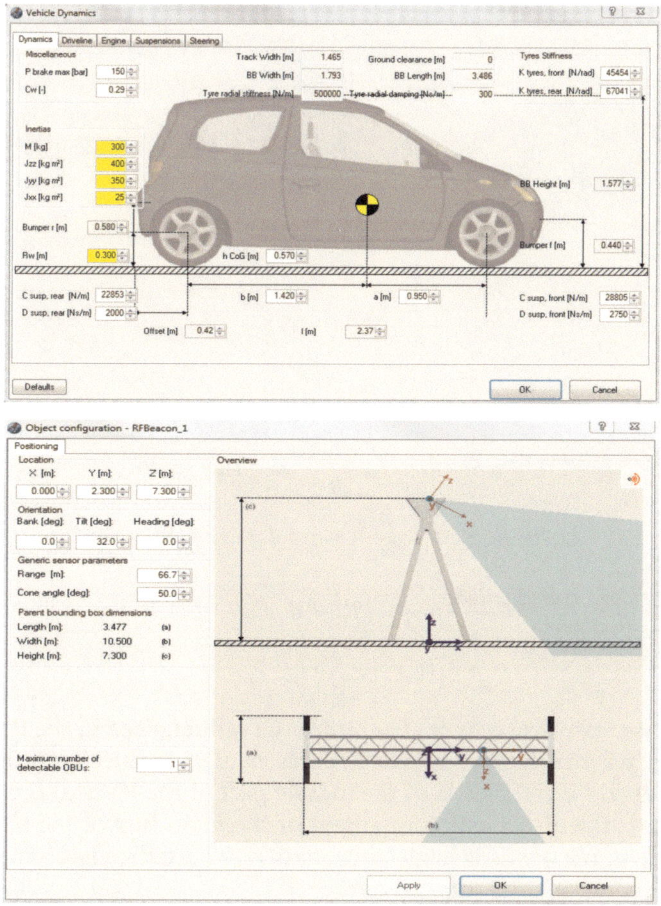

Figure 13.
Vehicle dynamics in PreScan *software.*

5.2 Dynamics designing

The elaboration of the control algorithms to model the dynamics and component structures of the vehicle takes place in *Simulink*, see **Figure 14**. Keeping interest in testing an electric vehicle, it was necessary to replace the default engine model, employed in *Prescan*, with an electric motor. The synchronous motor with permanent magnets was used (actually, an inverted structure, with the outer rotor, also called as an in-wheel motor). A considerable benefit of choosing this structure is the lack of mechanical transmission, which means that reduced mechanical losses are generated by the propulsion system.

The comparison between **Figures 8** and **14** emphasize the propulsion replacing, moreover, removal of the mechanical transmission blocks and gear shift logic had a place.

The motor was modeled in the rotor synchronous rotating frame (*dq* frame), and it was vector controlled, [21, 22]. The model of the concerned machine, a permanent magnet synchronous motor (PMSM) with surface-mounted magnets, contains the voltage equations (function of the fluxes or the currents), the torque balance dynamic equation, and the electromagnetic torque.

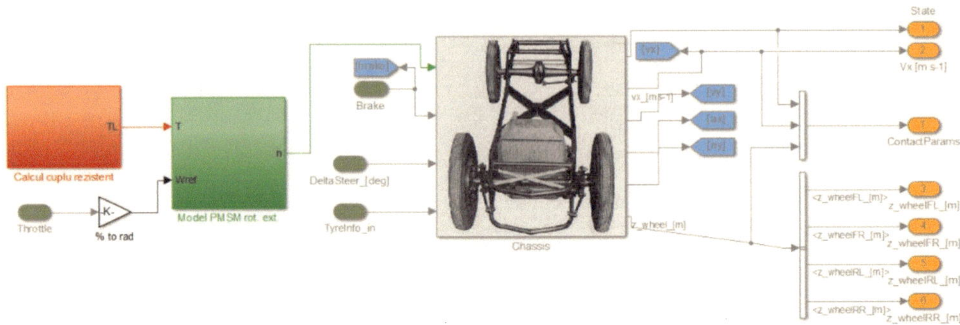

Figure 14.
EV dynamic model.

$$
\begin{aligned}
U_d &= R \cdot I_d + L_d \frac{dI_d}{dt} - \omega_r \cdot L_q \cdot I_q \\
U_q &= R \cdot I_q + L_q \frac{dI_q}{dt} + \omega_r \cdot L_d \cdot I_d + \Psi_{md} \cdot \omega_r
\end{aligned} \tag{1}
$$

$$
\frac{d\Omega_r}{dt} = \frac{1}{J_r} \cdot (T_e - B \cdot \Omega_r - T_r) \tag{2}
$$

$$
T_e = p \cdot \left[(L_d - L_q) \cdot I_d \cdot I_q + \Psi_{md} \cdot I_q \right] \tag{3}
$$

where R phase stator resistance, L_d, L_q are the inductances in the direct and quadrature axis, p is the number of pair of poles, ω_r, Ω_r are the electric and mechanical speed, respectively, J is the inertia of the mobile part of the PMSM, B is the air friction coefficient, Ψ_{md} is the direct axis component of the main flux of the permanent magnet (the one in the q axis is neglected), and T_e, T_r are the electromagnetic and resistive torque components, respectively.

In the *dq* coordinates frame, the supply of the PMSM is assured via an equivalent chopper which can offer two possible voltage values, $\pm U/2$, (U being the *dc* bus voltage), which means that the voltage reference vectors, u_d, u_q, are:

$$
\begin{bmatrix} u_d \\ u_q \end{bmatrix} = \frac{U}{2} \cdot \begin{bmatrix} u_d^* \\ u_q^* \end{bmatrix}, \tag{4}
$$

The control logic relies on PI controllers which will offer the quadrature current reference function of the speed error, the direct and quadrature axis reference voltages function of similar currents error, while the reference direct current is imposed to zero for maximizing the torque.

The *Simulink* diagram of the motor is depicted in **Figure 15**.

The mechanical model of the traction system, which integrates the operation down or up on the hills, is presented next.

Resistant torque applied to the motor shaft resulted from the vehicle resistant forces, namely climbing force (F_h), aerodynamic drag force (F_d), wind resistive force (F_w) and rolling force (F_r), [23, 24].

$$
F_m = F_{acc} + F_h + F_d + F_w + F_r \tag{5}
$$

where F_m is the motor force deduced from its torque and wheel radius and F_{acc} is the acceleration force.

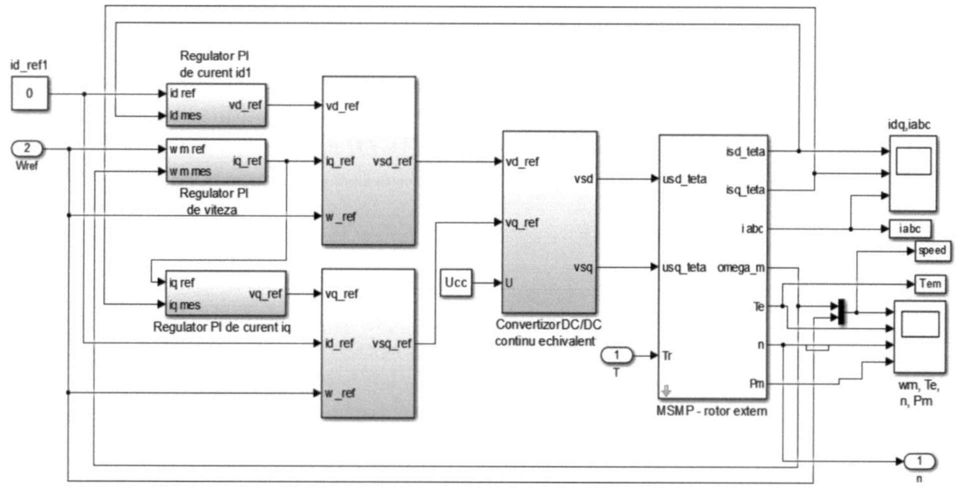

Figure 15.
PMSM model with a vector control technique.

The climbing force depends on the angle of incline θ, imported from VR, and uses the expression:

$$F_h = M_{tot} \cdot g \cdot sin\,(\theta) \tag{6}$$

where M_{tot} is vehicle mass (kg) and g is the gravitational acceleration (m/s^2).

The drag force depends very much on vehicle aerodynamics, considering its frontal area (A_f), air density (ϑ), aerodynamic coefficient (k_d) empirically determined for each body shape, and the vehicles speed (v), expressed as:

$$F_d = A_{fr} \cdot \vartheta \cdot v^2 \cdot k_d \tag{7}$$

In what concerns the resistance force due to wind, the wind relative coefficient (k_{rv}), the drag force, the vehicle's speed, and wind speed (v_w) are part of the equation:

$$F_w = \left(\left(0.98 \cdot \left(\frac{v_w}{v}\right)^2 + 0.63 \cdot \left(\frac{v_w}{v}\right)\right) \cdot k_{rw} - 0.4 \cdot \left(\frac{v_w}{v}\right)\right) \cdot F_d \tag{8}$$

The last resistant force, rolling resistance, is proportional to the road's surface coefficient (k_r) and angle:

$$F_r = k_r \cdot M_{tot} \cdot g \cdot \cos\,(\theta) \tag{9}$$

5.3 Communication designing

Depending on the hardware set-up, in some cases, the link between the hardware components must be configured. In the first case (a) where is only one connection, between HID and PC, usually via USB cable, there is no need to configure it due to drivers supplied by the HID manufacturer. Nevertheless, for case b there are two new links involved, between the RT platform and HID, and between the RT platform and PC, respectively. First of them, where the speed and break reference are sent toward the RT platform, ordinarily the signal type is analog, coming from a potentiometer, which could be connected to analog inputs of the RT platform. The other one, between the RT platform and PC, where information from VR traffic is sent to the RT platform and speed response back to VR, could be done using CAN, Ethernet, and even Serial communication. In the end, in case of *c*, compared to the previous case, an additional connection is defined between the RT platform and power train (PT). Again, this could be facilitated through CAN, Ethernet, Analog Signal, depending on the available hardware features.

6. Case study

In this section, the results obtained from several co-simulations will be presented. In all cases, the virtual vehicle was driven for one lap over the circuit shown in the map from **Figure 12**. Three software configuration variants and one HiL configuration are presented. These configurations were: *Prescan-Simulink*, *Prescan-Simulink-Amesim*, *Prescan-Simulink-dSpace,* and *Prescan-Simulink-dSpace*-PT. For cases where the motor has been simulated, the parameters are equivalent to the ones from the actual motor. Additionally, a V2I communication was tested through an RF sensor mounted on a traffic sign. Its purpose was to limit the speed of the vehicle when it was getting close to the first intersection.

The parameters of the electrical power train used in the simulation, and for the control of the real motor, when a specific battery voltage was used, are given in **Table 1**.

6.1 *Prescan-Simulink* co-simulation

Here we had a co-simulation that lasted for approximately 160 seconds, with results depicted in **Figure 16**. The speed reference provided by the throttle pedal, scaled in the motor-rated speed (0–44 rad/s) is observed, which is thoroughly followed by the motor, except for the moments of acceleration and deceleration, when, as expected, the error occurs due to vehicle inertia.

In the 20th second of the simulation, the V2I communication intervenes, which caps the speed at 30 km/h. The graph with the representation of the throttle pedal highlights this aspect, see **Figure 16**, where in yellow there is the resultant reference speed resulted by the speed limit algorithm. The antagonistic interaction of these two values is obvious, having as an operational period the time interval in which the vehicle is within the range of the RF sensor and the speed surpasses the set limit.

A correlation between the component of the current on *q* axis (red) and the electromagnetic torque developed by the motor (blue) is observed in **Figure 17**; there

PMSM parameters			
Rated power	**3 [kW]**	**Permanent magnet flux**	**0.0518 [Wb]**
Speed	420 [rpm]	Motor inertia	0.2214 [kg•m^2]
Rated torque	68 [Nm]	Mechanical friction coefficient	0.01
Pole pairs	28	Stator resistance	0.04 [Ω]
Supply voltage	48 [V]	*d*-axis inductance	0.4 [mH]
Rated current	63 [A]	*q*-axis inductance	0.4 [mH]

Table 1.
Real and simulated motor parameters.

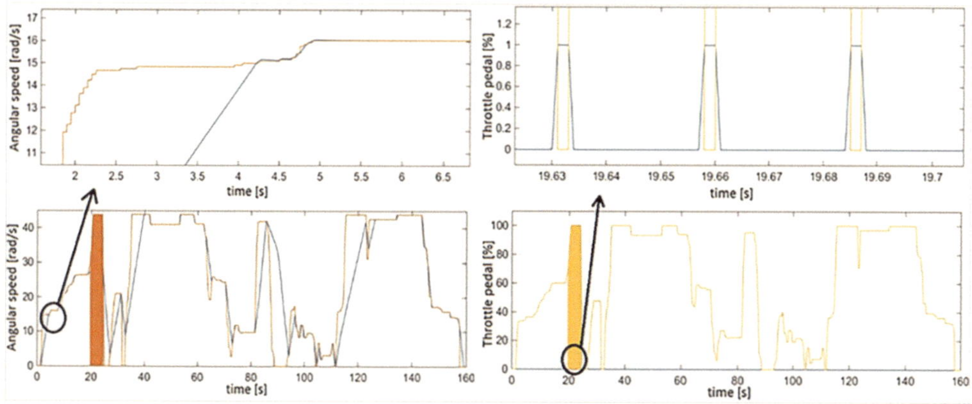

Figure 16.
Motor speed reference (red) and its response (blue).

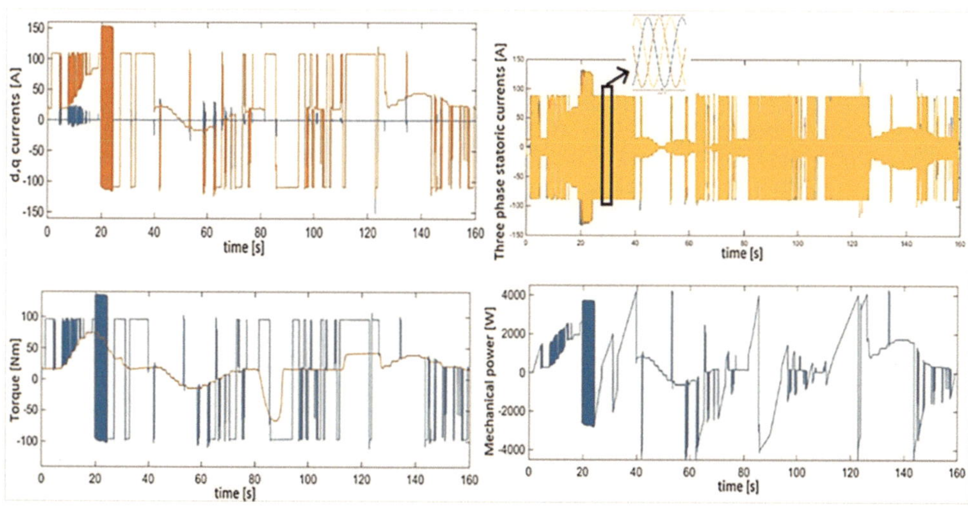

Figure 17.
Motor parameters evolution during simulation: Dq *axes currents, three-phase currents, torque, and mechanical power.*

is direct proportionality relation between them. For the acceleration and deceleration phases, the electromagnetic torque differs from the resistant one, which emphasizes the inertia presence. Of course, at the same time the current amplitude of the three-phase increases during these periods. Although the constructive parameters of the motor power is 3 kW, this is exceeded in the moments of acceleration/deceleration, due to the threshold set higher than the nominal one in order to obtain better dynamic performances. This is allowed only if it is mentioned in the motor specifications and can be done for short periods of time.

6.2 *Prescan-Simulink-Amesim* co-simulation

In the case of this simulation (with the results depicted in **Figures 18** and **19**), traversing the perimeter of the campus took approximately 168 seconds. The control strategy modeled in *Amesim* differs from the one in *Simulink*: here, the torque control strategy is approached.

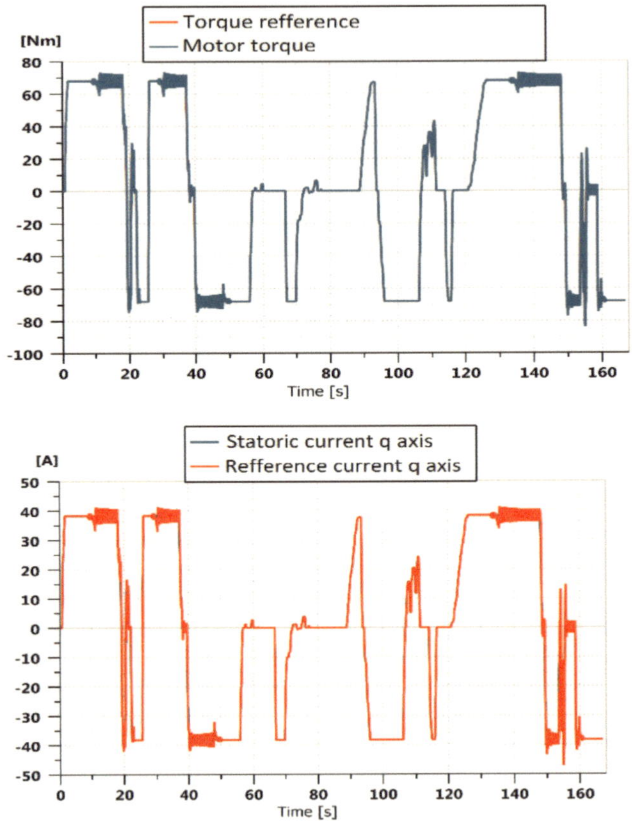

Figure 18.
Motor parameters evolution during: Torque and axis currents.

It should be mentioned that the flux weakening control strategy is also employed, the sub-model of the torque regulator being built to provide both references of the current, meaning the *d, q* components. The reference maintains the same source of origin as in the previous case, except that the scaling was done with torque values from 0 to 68 Nm, the maximum value of the rated torque of the machine. The reference torque is followed by the machine and due to a different control strategy, which limits the torque to a rated value, its value is less than half compared to previous results (*Simulink*). A reason for this change could be the different driving behavior of the driver.

The maximum angular speed of the motor exceeds the rated value, the reason being the strategy of flux weakening, possibly due to the value of the resistant torque, lower than the machine's rated one. The maximum linear speed of the vehicle in this cycle was approximately 18 m/s (64 km/h), with 5 m/s (18 km/h) more than in the previous case.

6.3 *Prescan-Simulink-dSpace* co-simulation

The results presented in this sub-section (see **Figure 20**) are dealing with a real-time system, based on dSpace platform (a MicroLabBox unit was used), in which the model from *Simulink* was deployed. The simulation time was about 230 seconds,

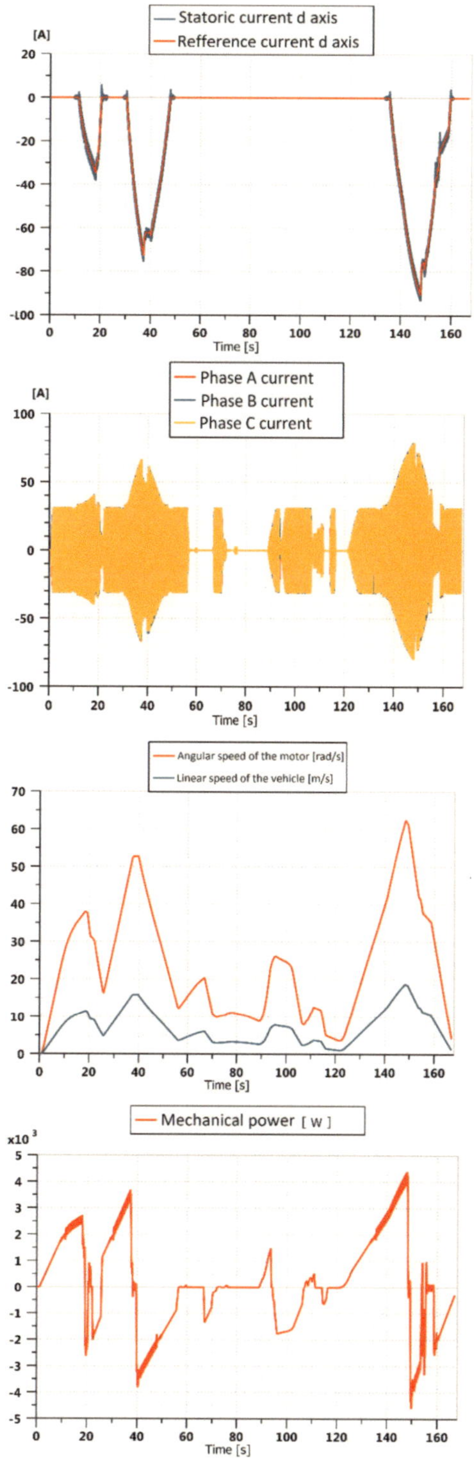

Figure 19.
Motor parameters evolution during simulation: Currents, speed, and mechanical power.

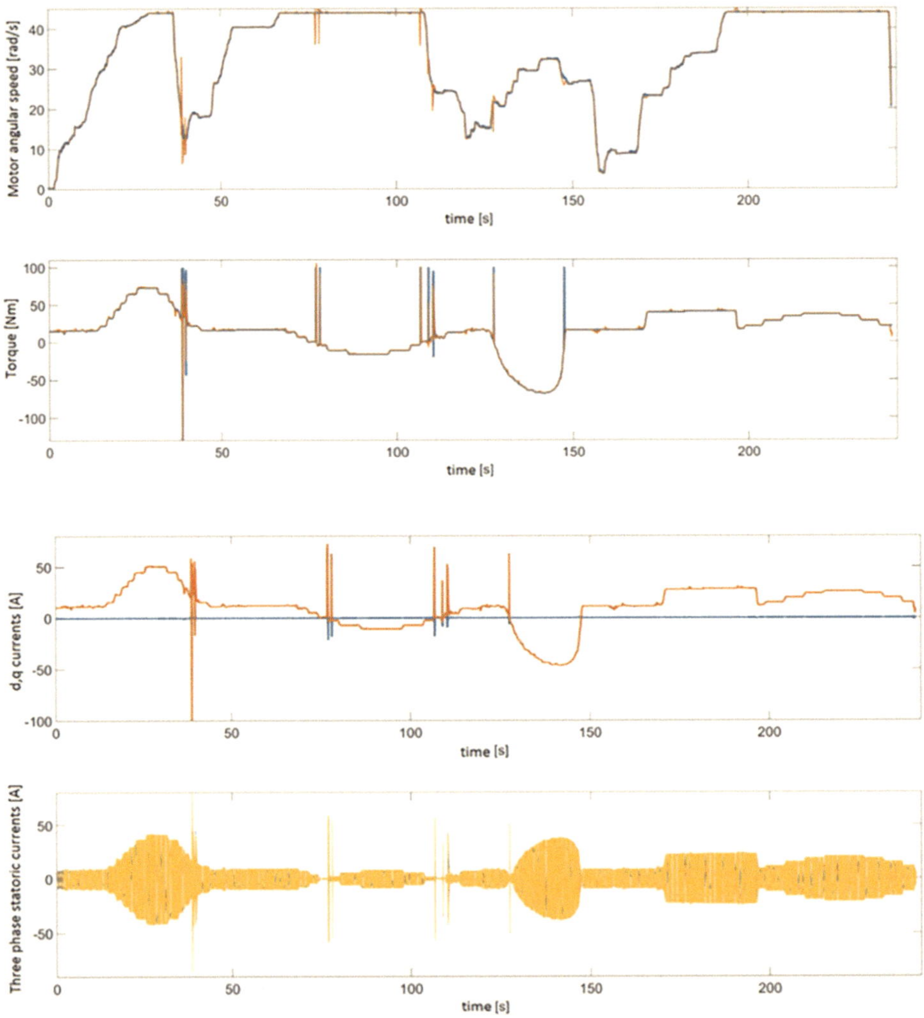

Figure 20.
Motor parameters evolution during simulation: Speed, torque, and currents.

so a less aggressive driving style was adopted. With the reduction of acceleration/deceleration, the value of dynamic torque was also reduced – an aspect visible in the torque plot.

From the representations of the three-phase currents and *iq* component, there is a decrease in amplitude, compared to the first results; this reconfirms a relaxed driving style. As in the case of the first presented results, the component element of the *id* current is zero, thus the flux weakening strategy remains absent.

6.4 *Prescan-Simulink*-dSpace-PT

In the last analyzed case, the exposed results are real (see **Figure 21**), measured in the virtual intermediate circuit and at the motor terminals, by means of transducers. The driving cycle lasted approximately 178 seconds, during which a moderate driving style was adopted.

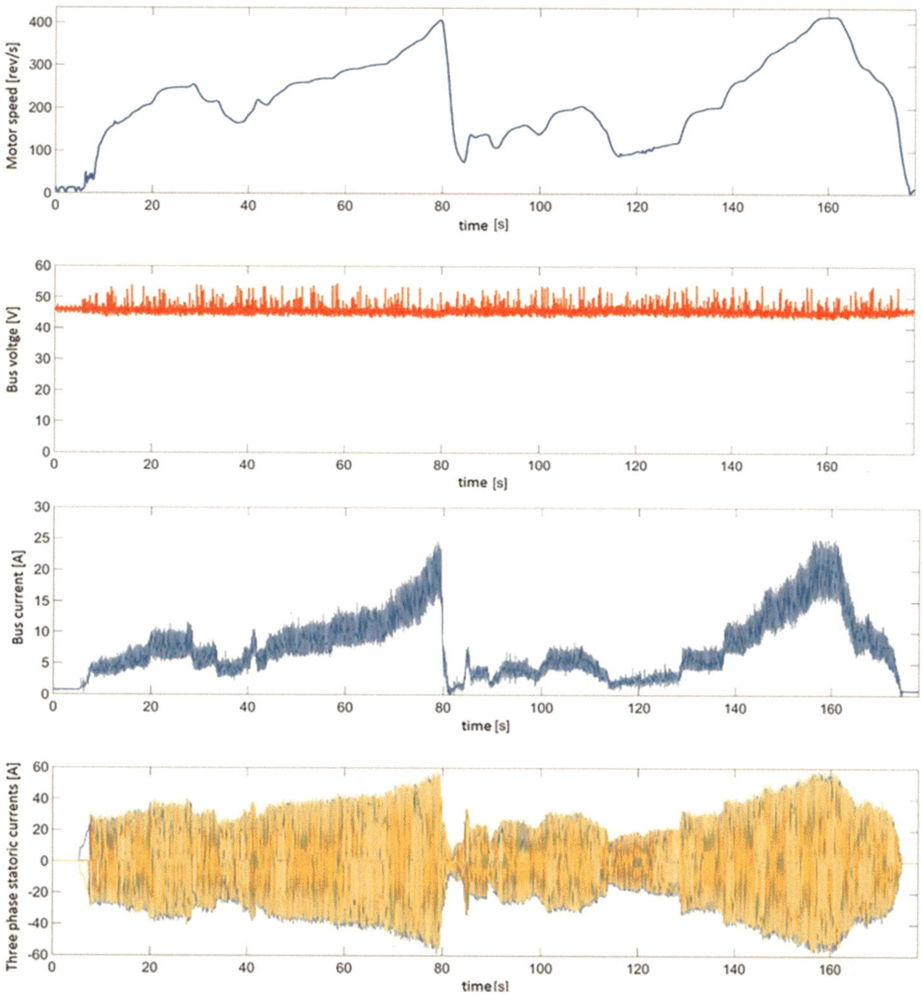

Figure 21.
Motor parameters evolution during HiL simulation: Speed, voltage, currents.

An attempt was made to replicate a driving profile similar to that in the previous simulations so that the obtained results could be comparable. Therefore, between the simulation results with the model employed on the RT platform and the real motor, at rated speed, it can be observed the approximate equal level for the amplitude of the stator currents. Additionally, the voltage and current flow from the battery were measured, where the presence of measurement errors is observed.

7. Conclusions

The need for vehicles with reduced or zero pollution potential, such as hybrid (HV), purely electric (EV), and fuel-cell (FCV) vehicles, pushes toward the evolution of their testing and validation, which implies that their progress follows an accelerated trend. In order to maintain this trend, a different approach for testing propulsion systems has been proposed here, by using driving simulators. This chapter focuses on the development of a car simulator that uses the virtual reality (VR) concept, with an online simulation model, being ranked at a realistic level of a fixed platform simulator.

The specialized software, embedded in the project, is represented by *Prescan*, *Matlab/Simulink,* and *Amesim* software packages, which are used for: modeling/simulating the VR (*Prescan*), modeling the vehicle dynamics (*Amesim* and *Simulink*), and its control (*Simulink*). The hardware components are also present, by using a personal computer (PC), a real-time (RT) platform (based on a dSpace board, i.e. a MicroLabBox unit), the graphical interface (special video-card and screens), and the human interface device (HID), meaning the steering wheel and pedals used by a "driver". The replication of the perimeter of the Technical University of Cluj-Napoca campus, inside the VR environment, was used as the test circuit, for different co-simulation configurations. A permanent magnet synchronous motor (PMSM) topology was tested on this circuit, both in the form of a mathematical model and a real motor. In both cases, the results are quite similar. In the simulated models, both control strategies, with speed and torque reference, were implemented and simulated and the expected results have been obtained. A simple vehicle control algorithm was also successfully tested.

The challenge of finding the right solution for the propulsion unit it is not an easy task, due to the vast number of mechanical and electrical constraints. The required performances of the electrical motor are dictated by the application, thus for an EV/HEV the most important mechanical performances are torque, speed, vibrations, and acoustic behavior because they affect the vehicle dynamics and the passengers' comfort. From the electrical point of view, the supply voltage, current, frequency, and efficiency are a few of the focus points, when the optimum operation is desired. The specific values for the mentioned parameters were emerged from the EV desired characteristics.

In conclusion, the results presented on the car simulator for testing the propulsion of electric vehicles, support and validate its use for the implementation, testing, and development of new hardware and software subsystems that are part of the electric propulsion system.

Acknowledgements

This work was supported by a grant of the Romanian Ministry of Research and Innovation, CCCDI - UEFISCDI, project number PN-III-P1-1.2-PCCDI-2017-0776/No. 36 PCCDI /15.03.2018, within PNCDI III.

Author details

Rares-Catalin Nacu* and Daniel Fodorean
Technical University of Cluj-Napoca, Romania

*Address all correspondence to: catalinnacu@emd.utcluj.ro

References

[1] European Environment Agency. Air quality in Europe – 2019 report. Available from: https://www.eea.europa.eu/publications/air-quality-in-europe-2019.

[2] European Commission, Directorate-General for Mobility and Transport, Next steps towards 'Vision Zero': EU road safety policy framework 2021-2030. Publications Office; 2020. DOI: 10.2832/391271

[3] Fodorean D, Idoumghar L. Motor variants for light EV optimized based on hybrid evolutionary algorithm. In: Proceedings of Electric Vehicles International Conference (EV); 3-4 October 2019. Bucharest, New York: IEEE; 2019. pp. 1-5

[4] Kim I, Kim H. Configuration analysis of plug-in hybrid systems using global optimization. In: IEEE Proceedings of World Electric Vehicle Symposium and Exhibition (EVS27); 17-20 November 2013. Barcelona, New York: IEEE; 2013. pp. 1-14

[5] Petit M, Marc N, Badin F, Mingant R., Sauvant-Moynot V. A tool for vehicle electrical storage system sizing and modelling for system simulation. In: IEEE Proceedings of Vehicle Power and Propulsion Conference (VPPC); 27-30 October 2014. Coimbra, New York: IEEE; 2014. pp. 1-5

[6] Sadoun R, Rizoug N, Bartholomeu P, Barbedette B, Le Moigne P. Influence of the drive cycles on the sizing of hybrid storage system battery-supercapacitor supplying an electric vehicle. In: Proceedings of 37th Annual Conference of the IEEE Industrial Electronics Society; 7-10 November 2011. Melbourne, New York: IEEE; 2011. pp. 4106-4112

[7] Zhou F, Lin H, Hu J, Lv Z, Qian B, Xu J. A novel balancing strategy for series-connected lithium batteries based on mixed charging mode. In: IEEE Proceedings of International Power Electronics and Application Conference and Exposition; 5-8 November 2014. Shanghai, New York: IEEE; 2015. pp. 732-736

[8] Guzella L, Sciarretta A. Vehicle Propulsion Systems. Springer; 2013

[9] Olstam J. Simulation of Surrounding Vehicles in Driving Simulators [Master's thesis]. Linköping Studies in Science and Technology; 2009

[10] Fisher DL, Rizzo M, Caird J, Lee JD. Handbook of Driving Simulation for Engineering, Medicine, and Psychology. CRC Press; 2011

[11] Engel A. Verification, Validation, and Testing of Engineered Systems. Wiley; 2010

[12] Daud ZHC, Asus Z, Bakar SAA, Husain NA, Mazali II, Chrenko D. Thermal characteristics of a lithium-ion battery used in a hybrid electric vehicle under various driving cycles. IET Electrical Systems in Transportation. 2020;**10**(3):243-248

[13] Fodorean D. Study of electric propulsion by using virtual reality software. In: IEEE Proceedings of 52nd International Universities Power Engineering Conference (UPEC); 28-31 August 2017. Heraklion, New York: IEEE; 2017. pp. 1-5

[14] Gros IC, Fodorean D, Marginean IC. FPGA Real-time implementation of a vector control scheme for a PMSM used to propel an electric scooter. In: Proceedings of the 5th IEEE International Symposium on Electrical and Electronics Engineering (ISEEE);

20-22 October 2017. Galati, New York: IEEE; 2017. pp. 1-5

[15] The Association of European Vehicle Logistics. WLTP, RDE and automotive emissions targets. 2019. Available from: https://ecgassociation.eu/wp-content/uploads/2019/10/19.10-WLTP-briefing-paper-v3-Draft-v1-Web.pdf [Accessed: 2022-12-17]

[16] Ligterink NE, Mensch P, Cuelenaere RFA, Hausberger S, Leitner D, Silberholz G. Correction algorithms for WLTP chassis dynamometer and coast-down testing. 2015. Available from: https://climate.ec.europa.eu/system/files/2016-11/wltp_correction_algorithms_en.pdf [Accessed: 2022-12-17]

[17] Tutuianu M, Bonnel P, Ciuffo B, Haniu T, Ichikawa N, Marotta A, et al. Development of the world-wide harmonized light duty test cycle (WLTC) and a possible pathway for its introduction in the European legislation. Transportation Research Part D: Transport and Environment. 2015;**40**: 61-75

[18] Karsolia A. Desktop Driving Simulator with Modular Vehicle Model and Scenario Specification [Master's Thesis, Chalmers University of Technology-Department of Applied Mechanics]. Sweden: Goteborg; 2014

[19] Kotilainen I. Driving Simulator Validity and Driver Behavior: Results of Driving Performance in Virtual Vs Natural Conditions. [Master's thesis]. University of Jyväskylä-Department of Computer Science and Information Systems; 2014

[20] Schwarz C, Gaspar J, Miller T, Yousefian R. The detection of drowsiness using a driver monitoring system. Traffic Injury Prevention. 2019; **20**(sup1):157-161

[21] Gyorgy T, Fodorean D. Model in the loop simulation of an electric propulsion system using virtual reality, In: IEEE Proceedings of 52nd International Universities Power Engineering Conference (UPEC); 28-31 August 2017. Heraklion, New York: IEEE; 2017. pp. 1-4

[22] Gyorgy T, Fodorean D. Human-in-the-loop simulation of an electric vehicle drivetrain. In: Proceedings of XIII IEEE International Conference on Electrical Machines (ICEM). 3-6 September 2018. Alexandroupoli, New York: IEEE; 2018. pp. 1545-1550

[23] Fodorean D. Global Design and Optimization of a Permanent Magnet Synchronous Machine used for Light Electric Vehicle. In: Book Chapter of 24 Pages in Monograph: Electric Vehicles – Modelling and Simulations, edited by Serif Soylu. London, UK, London, UK: Intech; 2011

[24] Irimia C, Grovu M, Husar C, Fodorean D, Antonya C. Co-simulation analysis for an electric vehicle powered by a high-speed electrical machine. In: IEEE Proceedings of Vehicle Power and Propulsion Conference (VPPC). 11-14 December 2017. Belfort, New York: IEEE; 2018. pp. 1-6

Chapter 4

Applications of Mixed Reality for Smart Aviation Industry: Opportunities and Challenges

Trung Hieu Tran, Yirui Jiang and Leon Williams

Abstract

Nowadays, mixed reality has improved operational efficiency and enhanced passenger experience in the aviation industry. Integrated with advanced machine learning and artificial intelligence techniques, mixed reality can easily deal with tons of aviation data to support decision-making processes in this industry. The chapter presents the state-of-the-art applications of mixed reality in smart aviation industry. Opportunities and challenges of integrating mixed reality with advanced machine learning and artificial intelligence techniques into the aviation industry are introduced. This chapter focuses on how the integrated mixed reality can improve the quality and reliability of maintenance, operation, piloting, training, and product design in smart aerospace engineering. It also describes autonomous, self-service, and data visualization systems in smart airports to enhance passenger experience. Finally, this chapter discusses airline's digital-based responses to the COVID-19 crisis.

Keywords: artificial intelligence, aerospace engineering, machine learning, mixed reality, passenger experience, smart aviation

1. Introduction

Along with the contribution of technology, data science (e.g., machine learning and artificial intelligence techniques) and mixed reality have brought much benefit into intelligent industrial revolution [1–3]. One of their significant impacts is to redefine the business models in intelligent aviation industry. In the aviation industry, mixed reality-based applications such as digital twin, aerospace design, manufacturing, verification, validation, and services have been widely implemented [4]. In addition, applications of smart visualization, mobility, robotics, and analytics were developed and integrated into the passenger journey [5].

Airlines have used mixed reality not only for efficiency improvement of staff operations but also for maximization of passenger satisfaction. For example, airlines are experimenting with mixed reality in smart maintenance in which repair works are transformed from manual to digital assistant to increase the efficiency of maintenance operation [6]. When applying mixed reality for training, virtual instructions are overlaid on physical items, allowing employees to gain hands-on experience [7]. For market, airlines use mixed reality for indoor navigation, advertising recommendations,

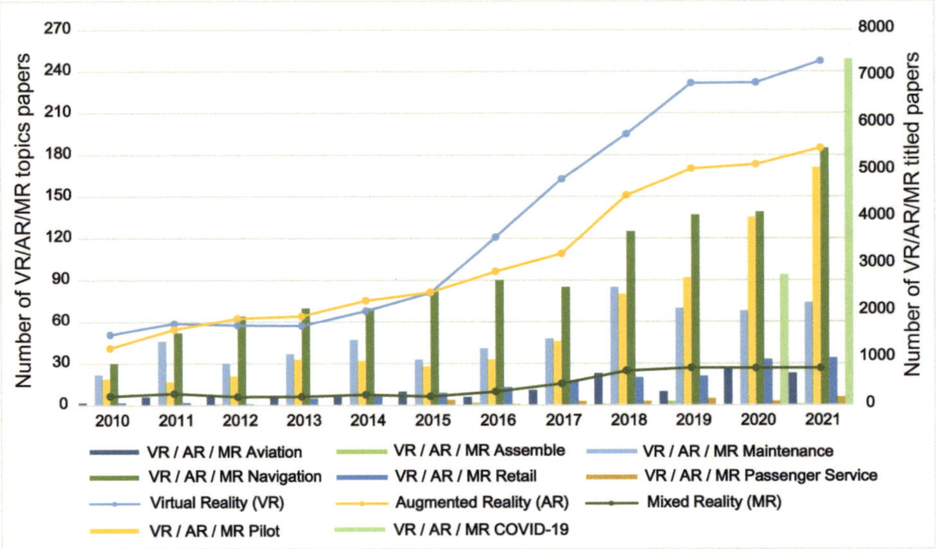

Figure 1.
The evolution of VR/AR/MR applications during the period of 2010–2021.

information notification, and immersive entertainment [8, 9]. Toward an automation industry, there is no doubt that mixed reality can transform the aviation industry by accelerating operator efficiency, developing staff skills, and improving the passenger experience.

The chapter provides a comprehensive review of existing works to understand the integrated mixed reality opportunities and challenges in the intelligent aviation industry. The keywords such as virtual reality (VR), augmented reality (AR), mixed reality (MR), VR/AR/MR aviation, VR/AR/MR design, VR/AR/MR assemble, VR/AR/MR maintenance, VR/AR/MR pilot, VR/AR/MR navigation, VR/AR/MR retail, VR/AR/MR passenger service, and VR/AR/MR COVID-19 were used to capture the most relevant literature published over the last decade from academic databases. Hundreds of papers that applied mixed reality for the aviation industry are then selected to review in the chapter. In addition to the opportunities and challenges of mixed reality in the smart aviation industry, the chapter also discusses a roadmap for intelligent aviation's future and the role of digital technologies in overcoming COVID-19 challenges. **Figure 1** shows that the number of papers published on VR/AR/MR has significantly increased over the last decade. It can be seen that the number of such papers has strongly developed since 2015 when the VR/AR/MR concept gradually received public attention, and its enormous potential for future development is widely acknowledged. Research works that proposed VR/AR/MR solutions to deal with the challenges of COVID-19 have also been concerned since 2020.

The rest of the chapter is structured as follows: Section 2 reviews MR-based aerospace engineering applications. Section 3 discusses how to apply MR to improve the passenger experience. Section 4 presents digitalization solutions implemented during and after COVID-19. Finally, Section 5 provides conclusions and future works.

2. Mixed reality for smart aerospace engineering

MR plays an integral role in the product design stage of manufacturing. Advanced interfaces and visualization systems are developed for product design, complex assembly, accurate maintenance, assisted piloting, and staff training. Engineers leverage MR to rapidly make improvements and changes before the product goes into manufacturing. Product designers overlay digital features onto the physical product and adjust it accordingly [10]. MR can thus shorten the development cycle and reduce the associated cost. For example, designers create a visualization of product hypotheses and prototypes and overlay virtual embodiment onto the real environment [11]. Engineers can interact with their designs using immersive tools to create a convincing product lifecycle [12]. The PJSC Scientific and Technical Council has recently created two VR sets and two mobile applications that received the highest rating and were recognized by aircraft simulator manufacturers, educational institutions, and aircraft operators [13].

Other application of MR in smart aerospace engineering is to assemble thousands of parts accurately and quickly [14]. Traditionally, manual instructions are painstakingly studied and constantly referred to as product assembly. Since the introduction of MR, detailed 3D instructions for hands-free and voice-controlled operations are provided to support product assembly efficiently [15]. Although installing electrical wires on an aircraft is an impossible task that requires zero error, interactive 3D wiring diagrams have made it possible [16, 17]. Airlines developed AR to provide technicians with hands-free, real-time 3D interactive wiring systems. Engineers' works have also been altered by MR. 3D manufacturing can improve 90% in first-time quality and reduce 30% in time cost than 2D manufacturing [18–20]. Gattullo et al. [21] reviewed 1757 papers to study the benefits of industrial augmented reality (IAR) in assembly, to demonstrate how MR can support manipulation, inspection, and positioning, and to build future research directions. Lai et al. [22] created a worker-centric system that uses Faster R-CNN in MR-based mechanical assembly. With CAD models as datasets, the authors successfully deployed the system for detecting simple tools. Hořejší et al. [23] compared MR-based instruction with traditional assembly one on 60 subjects. Staff turnover and proficiency can be improved by converting complex assembly manuals into on-screen virtual instructions. Alves et al. [24] compared the requirements, performance, and user satisfaction of various MR application products (i.e., handheld mobile AR, indirect AR that displays augmented scenes on a monitor, and see-through head-mounted displays) during assembly in a variety of industrial scenarios. Through 30 participants, user preference aspects, the advantages and disadvantages of each method, and the potential benefits of specific use cases were reported.

MR-based maintenance enables faster and more accurate operations [25, 26]. Mechanics have traditionally relied on paper manuals to assist with thorough maintenance. MR overlays instructions on top of the actual product, allowing operators to see machine status in real-time [27]. Quality screenings are essential for product maintenance. Identifying problems early on is beneficial for reducing machine downtime, and associated costs incurred by airlines [28]. On the other hand, human-based defect detection cannot guarantee optimal quality assurance because human errors are unavoidable. MR-based projects overlay over the product, allowing defects and inefficiencies to be identified with high accuracy and speed [29]. Maintenance technical documentation is becoming more visual. The most common maintenance operations in the manuals are integrated into remote maintenance support applications. Scurati et al. [27] created a glossary of graphic symbols to use in mixed reality. Mourtzis et al. [30] designed an AR-based remote maintenance framework that allows technicians and engineers to use real-time feedback by creating appropriate communication channels. The applicability of the developed framework was tested in the laboratory and real-world industrial scenarios with good performance. Verde et al. [31] combined the Internet of Things (IoT) and MR to create a dependable predictive maintenance system for off-site service. With IoT, engineers can share diagnostic information, and experts can monitor and guide maintenance operations remotely via the controller. Heterogeneous sensors integrated with AR visualization can improve operator safety in complex and dangerous plants. Liu et al. [32] proposed a multi-service collaborative machine intelligence predictive maintenance approach for fault prediction that combines CNN and LSTM. MR guides the visualization through the maintenance process, integrating massive amounts of data into the machine. Runji et al. [33] reviewed advanced MR research in manufacturing maintenance, focusing on analyzing user needs in applications via ergonomics, communication, situational awareness, intelligence sources, feedback, safety, motivation, and performance evaluation.

MR is used not only for navigation on the ground but also for assisting pilots [34]. MR creates visuals of terrain, navigation, air traffic, weather, instruments, and airspace information by overlaying information on demand. The MR-based application provides situational awareness in a user-friendly 3D format. Furthermore, a head mount display (HMD) guides pilots during the flight, takeoff, and landing. The MR systems show a corridor overlay to display pilots the proper path, ensuring their safety from takeoff to landing. The air-traffic visualization aids pilots in keeping an eye on the traffic and so increases passenger safety [35, 36]. Extensive training is required for airline personnel before they work in a real-world environment. However, preset manual training is insufficient to meet operators' growing needs. MR is thus developed to support training operators. It enables them to gain practical experience. When training and execution are combined, the equipment is easier to learn and use [37, 38]. Remote MR assists in lowering training and execution costs. It gives experts a view through the technicians' eyes, allows remote expert support, and performs inspections without regard for distance. In addition, remote MR makes the support process highly convenient and significantly reduces travel costs [39, 40]. Han et al. [41] conducted a systematic review of MR-based professional training applications to assess their overall effectiveness. Kaplan et al. explored the effectiveness of VR, AR, and MR for training through a meta-analysis and identifies that MR-based training can achieve as similar results as the traditional methods, but with less than time and cost. Moesl et al. [42] used MR for advanced pilot training in turbine engine aircraft and detailed the most challenging elements of the course that can be addressed with AR. A survey of 31 pilots and 22 instructors was conducted to evaluate MR-based pilot training tools thoroughly. To help pilots learn new flight panel layouts for different aircraft, Füchter et al. [43] developed MR as a human-machine interface for training. The system employs Human Centered Design (HCD), integrating a multi-disciplinary process involving flight instructors, students, and pilots, and has been tested in Brazil with positive user feedback. These MR applications show that it has the potential to transform aerospace engineering procedures toward reliant automation in the future.

3. Mixed reality for passenger experience enhancement

At the present, airports have implemented several initiatives to accelerate the development of commercial MR services in the aviation industry. HMD provides passengers with customized audio-visual options as the virtual entertainment market expands. Indoor navigation and sign guides based on MR technology significantly improve passenger travel efficiency. Passengers can see advertisements and targeted recommendations while walking. The virtual tags allow for faster baggage claims and reduce mishandled baggage. Real-time live dashboards display waiting times, flight arrangements, and weather information for passengers. The use of boarding reminders and real-time notification systems reduces the number of missed flight events and makes passengers feel more at ease at airports. Such MR-based initiatives demonstrate that the era of MR applications for improving passenger experience has arrived. **Figure 2** illustrates how airports utilize digital and MR to improve passenger experience.

Traditionally, airports installed beacons in terminals to provide passengers indoor navigation services that help new passengers navigate easily. However, this method is unreliable [44]. Recently, AR wayfinding displays directional prompts by superimposing digital information onto a real-world setting. It is a visual treat that makes life easier for visually impaired passengers by improving turn-by-turn audio prompts. An MR navigation system can locate late-running passengers and send them text reminders [45]. An AR-powered navigation app can even display points of interest along the way. The merchants send promotional messages to nearby passengers, informing them of current ongoing promotions. The app collects passenger information to create "people density" map in different areas that support airport queuing management by directing passengers to a less congested area. Passengers can use MR navigation to get through the airport quickly, accurately find their way, and avoid missing their flight. At the same time, it improves passenger flow management efficiency, promotes consumption, and increases airport and retail store revenue. [46] investigated the application of MR-based indoor positioning technology in airports and train stations. ARBIN was created to help passengers navigate through a user interface and navigation commands. The system was tested on nearly 1800 m^2 in 35 destinations and points of interest. The system performed very well and proved to be a practical indoor navigation solution for

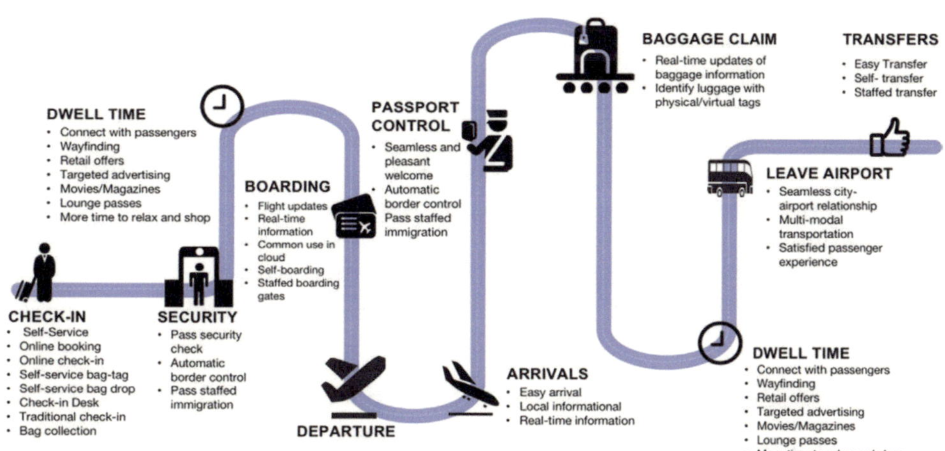

Figure 2.
The application of digital and MR to the entire airport passenger journey.

large buildings. Yoon et al. [47] made it easier for visually impaired people to navigate in indoor environments. The app "Clew" provides intuitive guidance along a route by recording the user's previous route and then using haptics and voice. Ayyanchira et al. [48] developed a collaborative, cross-platform, real-time MR-based navigation system. The system uses visualization and interactive methods to support common tasks in buildings through device sensors and image markers. Gladston and Duraisamy [49] developed an indoor navigation solution based on computer vision and MR to assist navigation for users in any new or unknown environment.

Retail contributes significantly to airport revenue. MR retail boosts in-store and online sales and alters customer purchasing habits. Customers, for example, can make better decisions with virtual try-on for wearable products, eliminating the need to measure and verify dimensions [50]. Nike customers are able to try on different colors and finishes in an immersive environment. It makes shopping more exciting and enjoyable. MR retail increases customer satisfaction while decreasing product returns. AR in retail has effectively converted potential leads by increasing engagement. The use of MR features allows the audience to interact with products. Customers can scan AR markers on print advertisements to access interactive augmented videos rather than plain old print ads. In addition, AR wayfinding can be combined with retail advertisements. Airports, for example, can show passengers the best shops along their route. Airports direct passengers through the airport based on their preferences and previous purchasing habits. Airports, for example, provide information on various product prices and guide customers to the correct location with guided arrows. According to recent research, MR is inspiring new traveler engagement. Riar et al. [51] presented a systematic literature review of MR retailing, analyzing technical features, user behavior, and experience. Tan et al. [52] outlined four broad usages of MR in the retail environment, assessing the potential of such applications and empirically examining their impact on sales. Janssen [53] tested MR-based shopping app as a solution for retailers to engage consumers and enhance their shopping experience. The study investigates consumer acceptance of mobile AR (MAR) shopping apps using the Technology Acceptance Model (TAM) and potentially external factors.

Airport incidents such as lost baggage or missed flights have received much attention by passengers. Implementing MR in airports has not only improved customer experience but also been beneficial in generating a return on investment (ROI). Before passengers boarding, airports use MR systems to notify them of the best boarding time and direct them to the right gate. Gatwick Airport deployed AR to provide personalized flight updates to travelers with a frictionless experience. Despite the importance of baggage in the passenger journey, baggage operations in airports continue to have many flaws [54, 55]. Many airlines are working to improve their baggage handling procedures. Singapore's Changi Airport uses smart glasses for baggage handling. Staff can instantly scan QR codes on baggage and cargo containers to obtain weight and unit data. MR has reduced the loading time from 60 to 45 minutes. MR has really increased the efficiency of airport operations, delivered more streamlined experiences, and, ultimately, provided more enjoyment for passengers.

4. Opportunities and challenges during and post COVID-19

Digital technologies are critical in assisting airports' recovery worldwide following the COVID-19 crisis. Most airlines are undergoing a digital transformation, resulting in cost savings, increased productivity, automation, streamlined operations, and

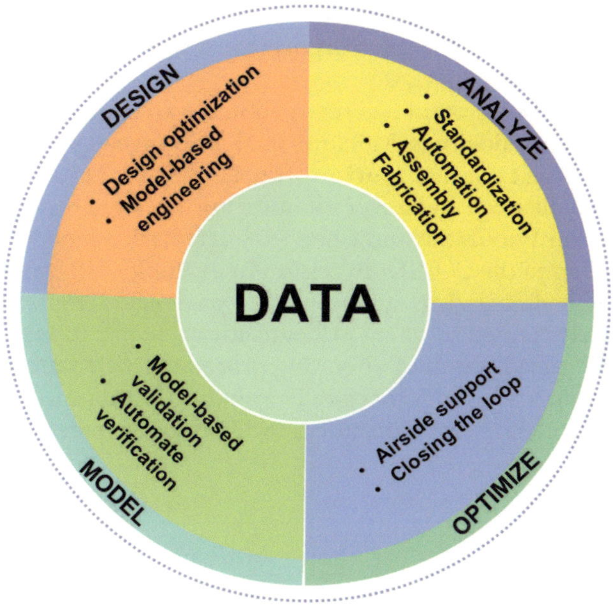

Figure 3.
The application of digitalization in the aviation industry.

reduced risk to smart airports [56]. **Figure 3** shows how to fully utilize data to drive the digital transformation of the aviation industry after COVID-19.

The length of time passengers spend at airports is critical in reducing the possibility of pandemic infections. Airlines are attempting to achieve seamless, touchless, and frictionless travel [57, 58]. Online ticketing, seat selection, and check-in reduce contact and queuing time. Automated check-in and boarding, self-service systems such as biometric enrollment, facial recognition, and finger scanning are applied. To reduce passenger anxiety, airports deploy notification systems to communicate with passengers and provide real-time traffic information. The airports can use mobile app to optimize queuing management and maintain a safe distance between passengers.

During the COVID-19 crisis, health screening and certificates are special operations [59]. To assist customer service representatives, airports use the web, mobile, social media, kiosk, email, smartphone apps, and chatbots. Each passenger's health is digitally recorded, and machine learning is used for abnormal data identification, identifying infected suspects in real time. Real-time data analytics ensure that potential patients are not overlooked. In addition, data science manages airport sanitization, ensuring high hygiene standards, which decreased human intervention and increased fidelity, accuracy, and observations.

Although airports will incur short-term losses during COVID-19, renewed digital aviation investments will increase long-term benefits [60]. It is not only necessary to promote airport digitization, but also to prioritize environmental protection [61, 62]. Passengers' modes of transportation to and from the airport have significantly changed in recent years. Airports need to prepare for an increase in passenger traffic in the future and a faster recovery of the leisure passenger segment [63]. Public airport transportation necessitates substantial support, such as route subsidies [64, 65]. Commercial air taxi services have been expanded at some airports. The Volocopter completed the first human-crewed flight over Singapore. Boeing, Bell,

Embraer, Safran, Uber, Fraport, and Groupe ADP, among others, have announced plans for urban air mobility solutions. Airports are working together to reduce their environmental impact.

An innovative travel ecosystem has emerged in the air transport industry. Based on WHO and UNWTO guidelines, Mohanty et al. [66] investigated the possibility of using MR to support the restart of tourism after COVID-19. The study explored the role of MR for social distance and lower mobility norms based on a systematic review of secondary data. Kerdvibulvech and Chen [67] summarized models and tools that use MR in responding to the COVID-19 crisis. For example, MR-based thermal imaging glasses are used to detect virus symptoms and passenger temperatures. Airport personnel is remotely trained by using MR educational applications. Priyan et al. [68] designed a MR-based standard operating procedure (SOP) for detecting and controlling social distance on a large scale. The application aims to help people control physical distance, accelerate social distance, and flatten the increasing case curve.

5. Conclusion

This chapter did a comprehensive review and provided a road map for MR applications in the intelligent aviation industry. Opportunities and challenges for implementing advanced MR in aerospace design, manufacturing, testing, and service are presented. MR technology helps smarter product design, safer assisted piloting, more flexible manufacturing, and more efficient staff training. The ongoing development of MR significantly impacts the passenger experience in the aviation industry, including visualized real-time information, AR wayfinding, virtual baggage tags, and AR retail. Furthermore, this chapter analyzed and discussed opportunities and challenges that the aviation industry can face during and after the COVID-19 pandemic. With the breathtaking array of new data-enhanced technologies, the aviation industry will open a new era of "touchless, seamless, and secure" operations and services in which the importance of machine learning and artificial intelligence techniques are concerned.

Acknowledgements

The authors would like to thank the Editor and the anonymous reviewers for their valuable comments.

DOI: http://dx.doi.org/10.5772/intechopen.108798

Author details

Trung Hieu Tran*, Yirui Jiang and Leon Williams
Centre for Competitive Creative Design, School of Water, Energy and Environment, Cranfield University, Bedfordshire, UK

*Address all correspondence to: t.h.tran@cranfield.ac.uk

References

[1] Zhu L, Yu FR, Wang Y, Ning B, Tang T. Big data analytics in intelligent transportation systems: A survey. IEEE Transactions on Intelligent Transportation Systems. Apr 23 2018;**20**(1):383-398

[2] Siegel JE, Erb DC, Sarma SE. A survey of the connected vehicle landscape—architectures, enabling technologies, applications, and development areas. IEEE Transactions on Intelligent Transportation Systems. 4 Oct 2017;**19**(8):2391-2406

[3] Menouar H, Guvenc I, Akkaya K, Uluagac AS, Kadri A, Tuncer A. UAV-enabled intelligent transportation systems for the smart city: Applications and challenges. IEEE Communications Magazine. 13 Mar 2017;**55**(3):22-28

[4] Brunton SL, Nathan Kutz J, Manohar K, Aravkin AY, Morgansen K, Klemisch J, et al. Data-driven aerospace engineering: Reframing the industry with machine learning. AIAA Journal. Aug 2021;**59**(8):2820-2847

[5] De Crescenzio F, Fantini M, Persiani F, Di Stefano L, Azzari P, Salti S. Augmented reality for aircraft maintenance training and operations support. IEEE Computer Graphics and Applications. 2010 Dec 23;**31**(1):96-101

[6] Henderson SJ, Feiner SK. Augmented Reality for Maintenance and Repair (Armar). New York, USA: Columbia University, Department of Computer Science; 1 Aug 2007

[7] Eswaran M, Gulivindala AK, Kumar Inkulu A, Bahubalendruni MR. Challenges and opportunities on augmented reality-based guidance in product assembly and maintenance/repair perspective: A state of the art review. Expert Systems with Applications. 7 Oct 2022;**213**:118983

[8] Pucihar KČ, Coulton P. Exploring the evolution of mobile augmented reality for future entertainment systems. Computers in Entertainment (CIE). 2015 Jan 28;**11**(2):1-6

[9] Safi M, Chung J, Pradhan P. Review of augmented reality in aerospace industry. Aircraft Engineering and Aerospace Technology. 16 Jul 2019;**91**(9):1187-1194

[10] Shen Y, Ong SK, Nee AY. Augmented reality for collaborative product design and development. Design Studies. 2010 Mar 1;**31**(2):118-145

[11] Ng LX, Wang ZB, Ong SK, Nee AY. Integrated product design and assembly planning in an augmented reality environment. Assembly Automation. 23 Sep 2013;**33**(4):345-359

[12] Mourtzis D, Zogopoulos V, Vlachou E. Augmented reality supported product design towards industry 4.0: A teaching factory paradigm. Procedia Manufacturing. 1 Jan 2018;**23**:207-212

[13] Neretin ES, Kolokolnikov PA, Mitrofanov SY. Prospect for the application of augmented and virtual reality technologies in the design, production, operation of aircraft and training of aviation personnel. Journal of Physics: Conference Series. 1 Jun 2021;**1958**(1):012030. IOP Publishing

[14] Tang A, Owen C, Biocca F, Mou W. Comparative effectiveness of augmented reality in object assembly. In Proceedings of the SIGCHI Conference on Human Factors in Computing Systems. 5 April 2003. pp. 73-80

[15] Gavish N, Gutiérrez T, Webel S, Rodríguez J, Peveri M, Bockholt U, et al. Evaluating virtual reality and augmented reality training for industrial maintenance and assembly tasks. Interactive Learning Environments. 2 Nov 2015;**23**(6):778-798

[16] Yin X, Fan X, Wang L, Qiu K, Liu R. Augmented reality training system for aerospace product assembly process guidance and its application. Aeronautical Manufacturing Technology. 2018;**61**(1):48-53

[17] Wang X, Ong SK, Nee AY. A comprehensive survey of augmented reality assembly research. Advances in Manufacturing. Mar 2016;**4**(1):1-22

[18] Davies P, Sivich L. Augmented reality and other visualization technologies for manufacturing in boeing. SAE International Journal of Aerospace. 2011;**4**(2):1133-1139

[19] Barfield W, Caudell T. Boeing's wire bundle assembly project. In: Barfield W, Caudell T, editors. Fundamentals of Wearable Computers and Augmented Reality (1st ed.). Boca Raton: CRC Press; 1 Jan 2001. pp. 462-482

[20] Baird KM, Barfield W. Evaluating the effectiveness of augmented reality displays for a manual assembly task. Virtual Reality. 1999 Dec;**4**(4):250-259

[21] Gattullo M, Evangelista A, Uva AE, Fiorentino M, Gabbard JL. What, how, and why are visual assets used in industrial augmented reality? A systematic review and classification in maintenance, assembly, and training (from 1997 to 2019). IEEE Transactions on Visualization and Computer Graphics. 6 Aug 2020;**28**(2):1443-1456

[22] Lai ZH, Tao W, Leu MC, Yin Z. Smart augmented reality instructional system for mechanical assembly towards worker-centered intelligent manufacturing. Journal of Manufacturing Systems. 1 Apr 2020;**55**:69-81

[23] Hořejší P, Novikov K, Šimon M. A smart factory in a smart city: Virtual and augmented reality in a smart assembly line. IEEE Access. 14 May 2020;**8**:94330-94340

[24] Alves JB, Marques B, Ferreira C, Dias P, Santos BS. Comparing augmented reality visualization methods for assembly procedures. Virtual Reality. Mar 2022;**26**(1):235-248

[25] Palmarini R, Erkoyuncu JA, Roy R, Torabmostaedi H. A systematic review of augmented reality applications in maintenance. Robotics and Computer-Integrated Manufacturing. 1 Feb 2018;**49**: 215-228

[26] Hincapié M, Caponio A, Rios H, Mendívil EG. An introduction to augmented reality with applications in aeronautical maintenance. In 2011 13th International Conference on Transparent Optical Networks. 26 June 2011 (pp. 1-4). IEEE.

[27] Scurati GW, Gattullo M, Fiorentino M, Ferrise F, Bordegoni M, Uva AE. Converting maintenance actions into standard symbols for augmented reality applications in Industry 4.0. Computers in Industry. 1 Jun 2018;**98**:68-79

[28] Ceruti A, Marzocca P, Liverani A, Bil C. Maintenance in aeronautics in an Industry 4.0 context: The role of augmented reality and additive manufacturing. Journal of Computational Design and Engineering. 1 Oct 2019;**6**(4):516-526

[29] Eschen H, Kötter T, Rodeck R, Harnisch M, Schüppstuhl T. Augmented

and virtual reality for inspection and maintenance processes in the aviation industry. Procedia Manufacturing. 1 Jan 2018;**19**:156-163

[30] Mourtzis D, Siatras V, Angelopoulos J. Real-time remote maintenance support based on augmented reality (AR). Applied Sciences. 8 Mar 2020;**10**(5):1855

[31] Verde S, Marcon M, Milani S, Tubaro S. Advanced assistive maintenance based on augmented reality and 5G networking. Sensors. 14 Dec 2020;**20**(24):7157

[32] Liu C, Zhu H, Tang D, Nie Q, Zhou T, Wang L, et al. Probing an intelligent predictive maintenance approach with deep learning and augmented reality for machine tools in IoT-enabled manufacturing. Robotics and Computer-Integrated Manufacturing. 1 Oct 2022;**77**:102357

[33] Runji JM, Lee YJ, Chu CH. User requirements analysis on augmented reality-based maintenance in manufacturing. Journal of Computing and Information Science in Engineering. 1 Oct 2022;**22**(5):050901

[34] Schaffernak H, Moesl B, Vorraber W, Koglbauer IV. Potential augmented reality application areas for pilot education: An exploratory study. Education Sciences. 2020 Mar 25;**10**(4):86

[35] Zollmann S, Hoppe C, Langlotz T, Reitmayr G. Flyar: Augmented reality supported micro aerial vehicle navigation. IEEE Transactions on Visualization and Computer Graphics. 24 Mar 2014;**20**(4):560-568

[36] Wu H, Cai Z, Wang Y. Vison-based auxiliary navigation method using augmented reality for unmanned aerial vehicles. In IEEE 10th International Conference on Industrial Informatics IEEE; 25 July 2012. pp. 520-525

[37] Macchiarella ND, Liu D, Vincenzi DA. 18 Augmented reality as a means of job task training in aviation. In: Hancock PA, Vincenzi DA, Wise JA, Mouloua M, editors. Human Factors in Simulation and Training (1st ed.). Boca Raton: CRC Press; 17 Dec 2008. p. 333

[38] Kaplan AD, Cruit J, Endsley M, Beers SM, Sawyer BD, Hancock PA. The effects of virtual reality, augmented reality, and mixed reality as training enhancement methods: A meta-analysis. Human Factors. Jun 2021;**63**(4):706-726

[39] Utzig S, Kaps R, Azeem SM, Gerndt A. Augmented reality for remote collaboration in aircraft maintenance tasks. In 2019 IEEE Aerospace Conference. IEEE; 2 March 2019. pp. 1-10

[40] Schneider M, Rambach J, Stricker D. Augmented reality based on edge computing using the example of remote live support. In 2017 IEEE International Conference on Industrial Technology. IEEE; 22 March 2017. pp. 1277- 1282

[41] Han X, Chen Y, Feng Q, Luo H. Augmented reality in professional training: A review of the literature from 2001 to 2020. Applied Sciences. 19 Jan 2022;**12**(3):1024

[42] Moesl B, Schaffernak H, Vorraber W, Holy M, Herrele T, Braunstingl R, et al. Towards a more socially sustainable advanced pilot training by integrating wearable augmented reality devices. Sustainability. 15 Feb 2022;**14**(4):2220

[43] Füchter SK, Schlichting MS, Salazar G. Aeronautic pilot training and augmented reality. Acta Imeko. 30 Sep 2021;**10**(3):66-71

[44] Huang H, Gartner G. A survey of mobile indoor navigation systems. In: Gartner G, Ortag F, editors. Cartography

in Central and Eastern Europe (1st ed.). Lecture Notes in Geoinformation and Cartography. Berlin, Heidelberg: Springer; 2009. pp. 305-319

[45] Kim J, Jun H. Vision-based location positioning using augmented reality for indoor navigation. IEEE Transactions on Consumer Electronics. 7 Oct 2008;**54**(3):954-962

[46] Huang BC, Hsu J, Chu ET, Wu HM. Arbin: Augmented reality based indoor navigation system. Sensors. 17 Oct 2020;**20**(20):5890

[47] Yoon C, Louie R, Ryan J, Vu M, Bang H, Derksen W, Ruvolo P. Leveraging augmented reality to create apps for people with visual disabilities: A case study in indoor navigation. In The 21st International ACM SIGACCESS Conference on Computers and Accessibility. 24 October pp. 210-221

[48] Ayyanchira A, Mahfoud E, Wang W, Lu A. Toward cross-platform immersive visualization for indoor navigation and collaboration with augmented reality. Journal of Visualization. Jun 2022;**25**:1249-1266

[49] Gladston A, Duraisamy A. Augmented reality indoor navigation using handheld devices. International Journal of Virtual and Augmented Reality (IJVAR). 1 Jan 2019;**3**(1):1-7

[50] Bonetti F, Warnaby G, Quinn L. Augmented reality and virtual reality in physical and online retailing: A review, synthesis and research agenda. Augmented Reality and Virtual Reality. 2018;**1**:119-132

[51] Riar M, Korbel JJ, Xi N, Zarnekow R, Hamari J. The use of augmented reality in retail: A review of literature. In Proceedings of the 54th Hawaii International Conference on System Sciences 5 January 2021 p. 638

[52] Tan YC, Chandukala SR, Reddy SK. Augmented reality in retail and its impact on sales. Journal of Marketing. Jan 2022;**86**(1):48-66

[53] Janssen C. Consumer Acceptance of Mobile Augmented Reality Shopping Applications in Stationary Retail Trade. Mid Sweden University, Faculty of Human Sciences, Department of Business, Economics and Law, Sundsvall and Östersund, Sweden; 2018

[54] Zhang T, Ouyang Y, He Y. Traceable air baggage handling system based on RFID tags in the airport. Journal of Theoretical and Applied Electronic Commerce Research. Apr 2008;**3**(1):106-115

[55] Singh A, Meshram S, Gujar T, Wankhede PR. Baggage tracing and handling system using RFID and IoT for airports. In 2016 International Conference on Computing, Analytics and Security Trends. IEEE; 19 December 2016. pp. 466-470

[56] Serrano F, Kazda A. The future of airports post COVID-19. Journal of Air Transport Management. 1 Oct 2020;**89**:101900

[57] Albers S, Rundshagen V. European airlines' strategic responses to the COVID-19 pandemic. Journal of Air Transport Management. 2020 Aug 1;**87**:101863

[58] Gössling S. Risks, resilience, and pathways to sustainable aviation: A COVID-19 perspective. Journal of Air Transport Management. 1 Oct 2020;**89**:101933

[59] Gold L, Balal E, Horak T, Cheu RL, Mehmetoglu T, Gurbuz O. Health screening strategies for international air travelers during an epidemic or pandemic. Journal of Air Transport Management. 1 Mar 2019;**75**:27-38

[60] Adrienne N, Budd L, Ison S. Grounded aircraft: An airfield operations perspective of the challenges of resuming flights post COVID. Journal of Air Transport Management. 1 Oct 2020;**89**:101921

[61] Chaouk M, Pagliari R, Moxon R. The impact of national macro-environment exogenous variables on airport efficiency. Journal of Air Transport Management. 1 Jan 2020;**82**:101740

[62] Rolnick D, Donti PL, Kaack LH, Kochanski K, Lacoste A, Sankaran K, et al. Tackling climate change with machine learning. ACM Computing Surveys (CSUR). 8 Feb 2022;**55**(2):1-96

[63] Schultz M, Evler J, Asadi E, Preis H, Fricke H, Wu CL. Future aircraft turnaround operations considering post-pandemic requirements. Journal of Air Transport Management. 1 Oct 2020;**89**:101886

[64] Mandle PB, Mansel DM, Coogan MA. Use of public transportation by airport passengers. Transportation Research Record. 2000;**1703**(1):83-89

[65] Vuchic VR. Urban public transportation systems. Philadelphia, PA, USA: University of Pennsylvania; 5 Jan 2002;**5**(5):2532-2558

[66] Mohanty P, Hassan A, Ekis E. Augmented reality for relaunching tourism post-COVID-19: Socially distant, virtually connected. Worldwide Hospitality and Tourism Themes. 2 Dec 2020;**12**(2):753-760

[67] Kerdvibulvech C, Chen LL. The power of augmented reality and artificial intelligence during the COVID-19 outbreak. In International Conference on Human-Computer Interaction. Cham: Springer; 19 July 2020. pp. 467-476

[68] Priyan L, Johar MG, Alkawaz MH, Helmi RA. Augmented reality-based COVID-19 SOP compliance: Social distancing monitoring and reporting system based on IOT. In 2021 IEEE 12th Control and System Graduate Research Colloquium. IEEE; 7 August 2021. pp. 183- 188

Chapter 5

Learning Composition and Architectural Design with Immersive Virtual Reality Application CREALITY 1.0

Carlos Augusto Rengifo Espinosa
and Oton Alberto Navas de la Cruz

Abstract

This document focuses on identifying how the CREALITY 1.0 application (self-development) played an important role in teaching and teaching processes in the architectural career, working with topics such as architectural composition and supported by immersion in virtual realities that could stimulate student creativity in a college education setting. As a theoretical basis, this research and exploration exercise focused on the composition and formal principles underpinning the student-generated design proposals. The evaluative component of the exercise was directed toward the analysis of evidence of learning using the application, and it was possible to emphasize that the advantages in execution time, expenses of materials and dynamization in the cognitive processes allow considering the application as a relevant tool with the field of education in architecture and the present world.

Keywords: computer processes, digital mediation, volumetry, architecture, architectural design

1. Introduction

For university education in architecture, it is important to consider the demands of countries and the increasingly globalized world, not only in this area of professional practice; also, in relation to the formative profile that academic programs seek to develop in response to these challenges. In this sense, the prevailing need for education to generate knowledge is added, with teaching processes based on the support of information and communication technologies (ICTs) that can extend the boundaries of imagination easily and easily.

These educational boundaries were demarcated by traditional teaching, characterized by a development of processes that could only be supported by physical materiality. Now, immersive virtual reality is a key alternative within the educational realm, because it becomes a versatile method for learning by playing, doing and building [1]. The virtual realities offered by today's technologies allow to establish differences and interactions between the perceived real world and the subjective world, opening up multiple learning options or possibilities; What del Vasto [2] stands out when referring to educational processes centered in the development of creative and actitudinal skills. In the current medium of higher education, there are strategies and didactics, whose objective is to give shape to the architectural sensitivities of the students and to develop their communicative skills, as well as their problem solving [3]. According to some authors, the traditional character of the teaching of composition was characterized by being constituted from simple figures or forms, which acquired complexity when understood together [4]. This is a starting point toward evolution as a final architectural design or buildable model, configured by aesthetic elements, technological and creative aspects that will allow its habitability. For this reason, design as a process should focus on activities that develop the complete visualization and exploration of the components of the ongoing project [5]. Such a process must be based on "spatial ability as a component of human intelligence. It is understood as the ability to degenerate, retain, retrieve and transform well-structured visual images" ([6], p. 2). Based on the above, the new challenges of university education are oriented toward the application of new technologies that can provide technical aspects and facilitate the educational processes. For Escorcia and de Trivino [7], the use of information and communication tecnologìas are priorities that are established from the guidelines managed by the Ministry of Education in Colombia. This implementation should encompass every area of professional knowledge. This technology education will allow students to interact with devices on a daily basis in a normal way. Other authors argue that inclusive education with technology is required; instead, it must be supported by it [8].

2. Architectural composition, the developmental method and virtual reality

At the University of La Costa (Barranquilla, Colombia), a developmental pedagogical model is being implemented that is based essentially on processes of learning by doing [9]; and in the case of the architecture program, they are process activities that involve topics such as proportion, geometry or composition, including the study of the context where it can be implemented in a created project. Finally, these processes arrive at an architectural proposal with contextual bases that would allow their materialization and implementation. Virtuality as a teaching method demands versatile forms of work and pedagogies to dynamize the relationships generated between the precepts and concepts of architectural discipline, applied in the design proposal, in this case, from a virtual reality. This work has a starting point. From the basics of Ching [3], it corresponds to speaking of plane and volume. Celani [10], for its part, allows to define other complements and principles, fundamental in the teaching and learning process with virtual reality; these refer to balance, rhythm, pattern and proportion, in interaction with algorithms or protocols for its modeling. Geometry, for obvious reasons, plays an important role here. In these architectural design lessons, the instructor makes some recommendations and corrections about the student's project; this allows you to determine that architectural education is learned by trial-and-error method [11]. In addition, the education of the contemporary post-pandemic world allowed to dynamize virtualization scenarios, with which educational advances focused on the use of immersive technology to improve learning based on experiences that foster collaborative commitment and increased creativity;

therefore, "it is beneficial for students to add immersive experiences to the teaching process as appropriate" ([12], p. 4).

2.1 Methodology

The research process carried out had a mixed methodological basis; that is, a qualitative and quantitative nature [13–15]. Similarly, progressive levels of descriptive, experimental, and applied research aspects were managed. Procurations were involved among those resulting from abstract thought in articulation with some perceptible elements of reality. In combination, it was induced to a design process in which the subject (student), the concept (abstract reality), and the technology (virtual immersion) interacted. With a quasi-experimental research design, a group of subjects (architecture students) were targeted, who actively participated in the process and were involved in practical activities, based on systematic and pre-established observations, according to the stages of the research. The application of control surveys and observation sheets supported these follow-up activities, focusing on assessments of the performance of the CREALITY 1.0 software. The student population involved was 57 subjects in the fifth semester, aged between 18 and 27 years and belonging to the architecture program of the University of La Costa, Barranquilla, Colombia. This group was selected because the subject of the design subjects was adapted to the needs of the research. Basically, the sequence in the data collection and collection stage consisted of the students performing the design in virtual reality, in step that they executed surveys of control or follow-up of the application. As for the operational form of the software, it is important to mention that its use consisted of a virtual environment weightless that gives existence to pure forms, susceptible to transformation according to the ideas of design and composition. It presents two virtual hands that act as a user (student) simulator; as visually shown in **Figure 1**. The application has a collection of forms and solid bases very varied and sufficient. In addition, it presents other sound qualities of ambience and free tools or commands. The basis of this project was the use of the development

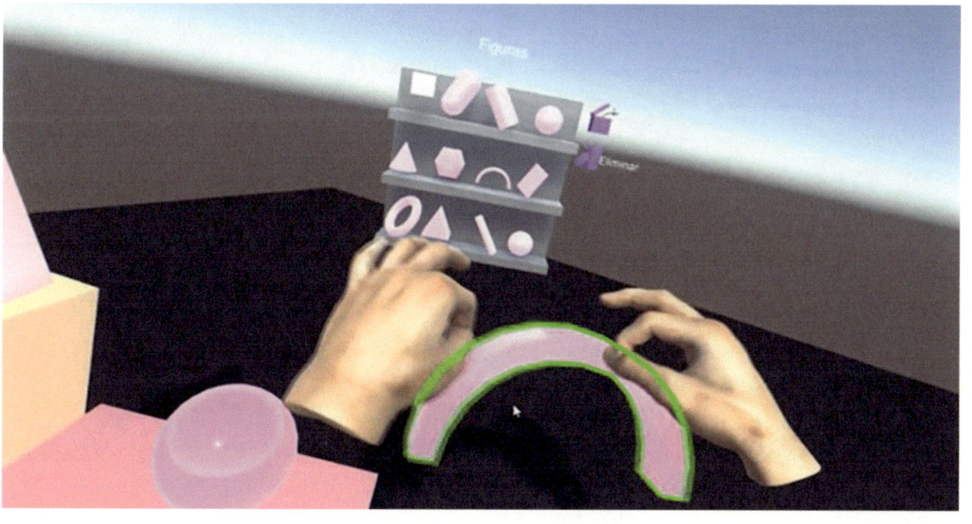

Figure 1.
CREALITY 1.0 overview.

engine called Unity V 2019.2.4, using the Leap Motion control interfaces: core, Hands Module and Interaction Engine. The visualization processes used the Oculus Rift DK2 hardware.

The haptic immersive reality sensors of CREALITY 1.0 enable the creation of objects (volumes or solids) for the configuration of volumetric compositions. The properties or transformations that the application allows are scaling, blends, displacements, rotations, specific locations, among other options. The practical stage of the research process with the students began with the request for a first sketch or proposal written by freehand and on paper. The participating group was then asked to prepare a first proposal worked on the virtual application, as shown in **Figure 2a** and **b**. Both workshops were held at the University of the Coast, with the support of physical computer educational resources. The designated time range was 45 minutes with observation and recording traces of the sessions. This data collection was carried out on the basis of questionnaires sent and completed by the participants. It was structured with 28 question items that were organized into four components: personal data, early stage of immersion virtual reality, creativity and finally, the early stage of the proposed design. The questions in the questionnaire were closed and completed via Outlook.

2.2 Operational variables

The operational variables of the research were three. An independent one (see **Table 1**), called "integration of immersive virtual reality"; and two dependent type variables, called "creativity" and "early stage of design" (see **Table 2**).

Based on the results consolidated by the application of the observation sheet and the questionnaires, it was possible to determine that the indicator of prior knowledge on virtual reality was rated at 1.64 out of 5.0, i.e., equivalent to 33% of previous knowledge in the group of participants. It was possible to infer that, although the students are native internet users and digital natives, they did not possess the basic

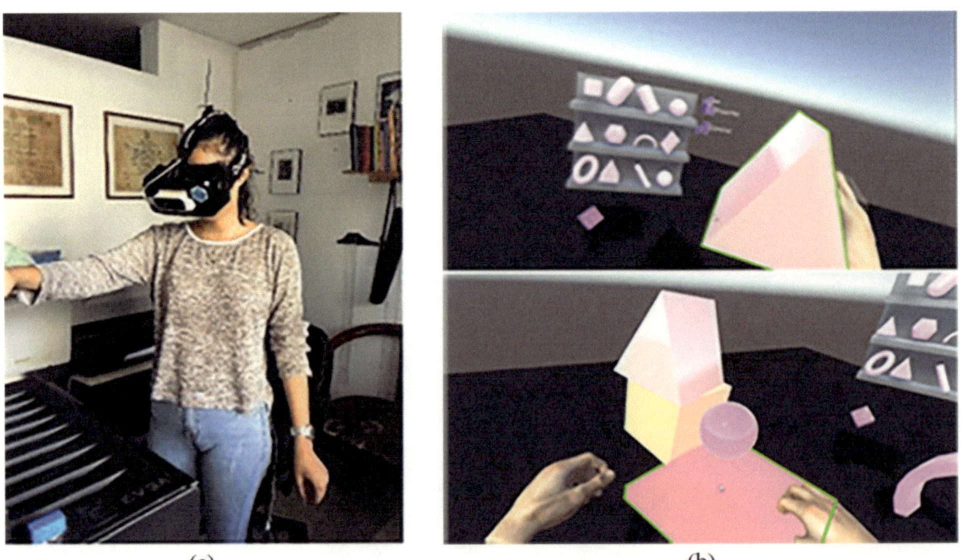

(a) (b)

Figure 2.
(a) Physical experimentation with hardware. (b) Virtual experimentation with hardware.

	Variable	Dimension	Indicator
Independent	Integration of immersive virtual reality (IR) in the early stage of architectural design. (This variable is intended to familiarize you with the tool, which will allow, through the use of the IR, the conception of your creative idea.).	Technique	Ease of use of the tool
			Appropriation of the tool
			Previous knowledge
		Methodology	Creation process
			Awareness of forms
			Resulting composition
		Attitude	Use of the tool
			Development of the exercise

Table 1.
Independent operational variable.

	Variable	Dimension	Indicator
Dependent	Creativity [16] (Seeks the student's sensitivity to a given problem that leads him/her to identify difficulties, speculate and modify in order to reach a compositional result.)	Fluency	Associativity
			Ideation
			Expression
		Flexibility	Spontaneity
			Perseverance
			Adaptability
		Originality	Statistics
			Remote actions
			Quality of responses
		Motivation and curiosity	Curiosity
		Elaboration	Elaboration
		Sensitivity to problems	Problem identification
			Problem solution
	Early design stage [4] (The synthesis of the composition that will allow to make the general decisions of what will be the architectural project.)	Compositional or perceptual axes	Generating frame
		Organization and significance of the form	Symbolic demand
		Shape Transformations	Generic and specific forms
		Articulations of the form	Connecting, connected, complementary and adjoining spaces
		Proportionalities of the form	Scale and modulus ratio
		Motivation and curiosity	Curiosity
		Elaboration	Preparation of the idea

Table 2.
Dependent operational variable.

theoretical knowledge for the exercise, which implies the need to sustain their classroom training, as shown in **Figure 3**.

The data obtained by the indicator of ease of creation showed an average of 4.22 of 5.0, corresponding to 84%. With this data, it was possible to determine that the digital tool strengthens the agility to learn in the participants (see **Figure 4**). Likewise, the second indicator of volume awareness and spatial location averaged 4.33 out of 5.0, thus 87% development. It was possible to determine that the participants are digital natives and learn intuitively through audio-visual sources, such as videos, video games, images, etc. The third indicator of the level of approach to the imagined composition obtained a 4.04 of 5.0, that is, 81%; showing the work with the representative coordinates that the participant has from his real experiences.

With **Figure 5** it was possible to identify that the dimensions of fluidity and flexibility obtained respective scores of 4.61 out of 5.0 equivalent to 92%: and 4.42 out of 5.0 equivalent to 88%; two values being high in this measurement. Both dimensions are related to the processes of participants' perception of reality. Another very high rating was that of curiosity and motivation with 4.49 out of 5.0, equivalent to 92% in the measurement.

The following **Figure 6** showed that the dimension of the compositional axes obtained a 3.35 of 5.0, equivalent to 70% and placed in a high measuring range, together with the dimensions of shape meaning, proportionality, balance, rhythm and pattern, and shape meaning. It was important to emphasize that the application offers freedom of work in respect of compositional management, with respect to the classical forms of Cartesian representation. The application encourages generative ways of creating architectural proposals. On the other hand, other measuring ranges were found close to an intermediate level, as is the case with the Shape Transformations dimension, with 2.92 out of 5.0 equivalent to 58.4%. This reflected some limitations of the application in this process of transformation in virtual forms.

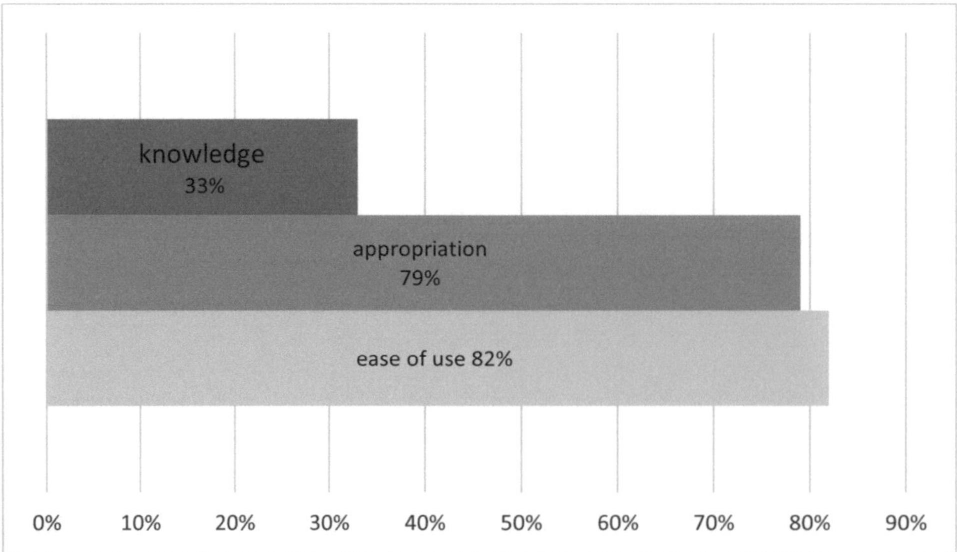

Figure 3.
Assessment of the dimension: methodology.

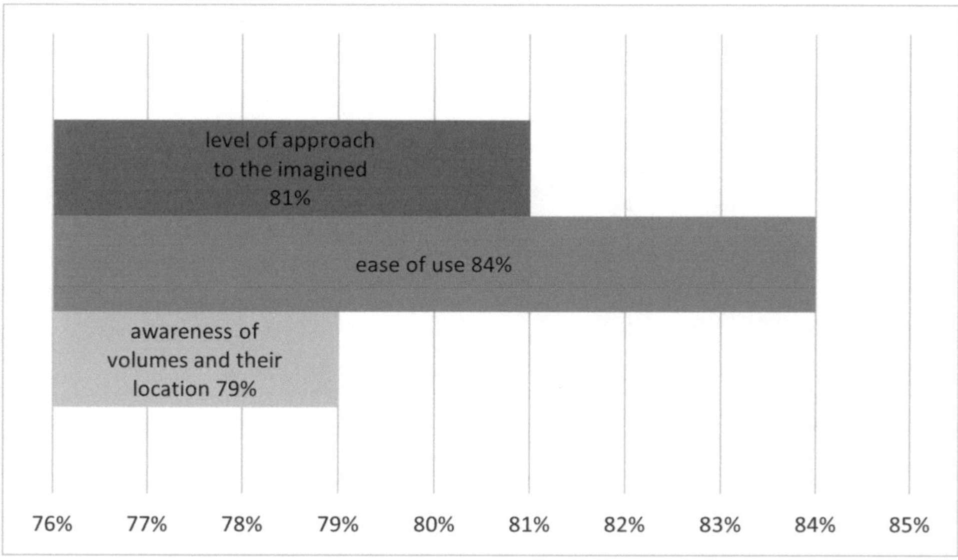

Figure 4.
Assessment of the variable: integration.

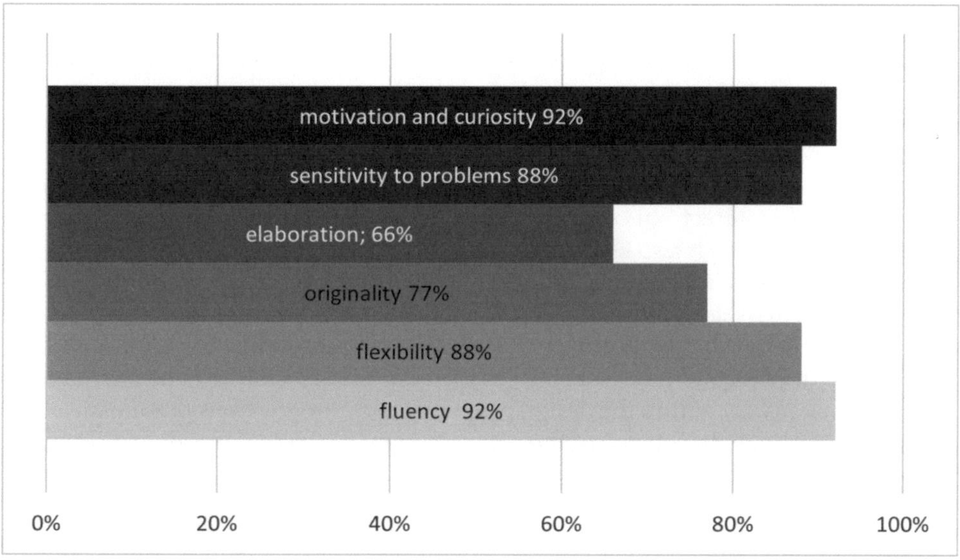

Figure 5.
Assessment of the variable: creativity.

The results obtained allowed to affirm that the application CREALITY 1.0 is a tool with high values of relevance, due to the percentages reached in the early stage of architectural composition. Its dimensions measured in the data described above show that the design process achieved by the participants was highly based on the compositional axes, meaning of the forms, transformation forms, articulation forms, proportionality, balance and rhythm and guideline.

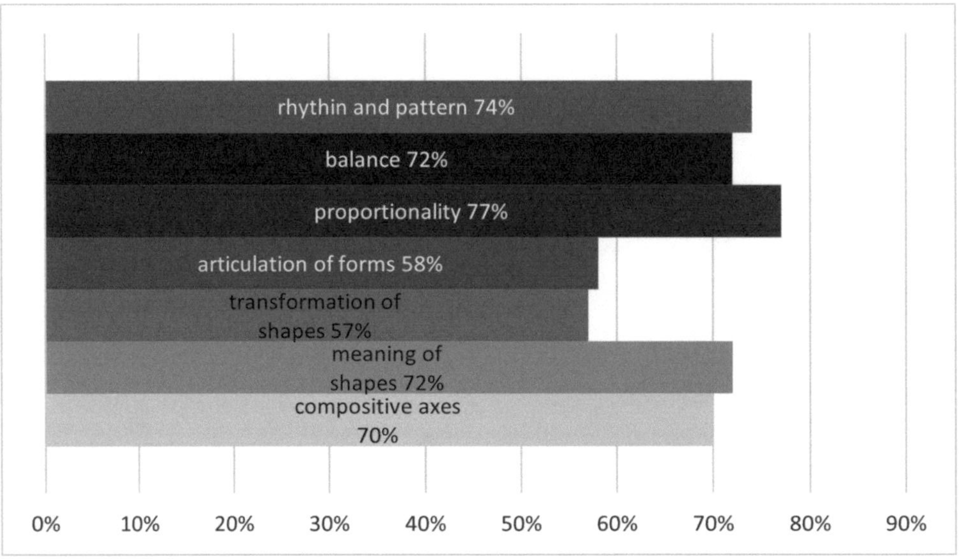

Figure 6.
Assessment of the variable: early design stage.

3. Conclusion

The central conclusions that were obtained from this research are concretized in affirming that CREALITY 1.0 is a pedagogical alternative that allowed an interactive experience with the architectural composition of analog and material form, in an educational scenario of the Architecture Program of the University of La Costa, Barranquilla (Colombia). This tool managed to generate interest on the part of the students toward the realization of architectural compositions and increased the possibility of generating alternatives in less time. It also became an incentive for the generation of volumetrics in the early stages and its linkage created links of greater interaction between the teacher and the students. The use of this strategy allowed to strengthen new teaching processes and to face another of the important challenges from the student's vision, because it allowed its appropriation in the working group and it was possible to generate remarkable levels of creativity within the design proposals and compositions presented. On the other hand, the central theme of the research ensured the prolonged continuation of both the descriptive and experimental phases, allowing better results to be obtained with future research related to immersion and virtual education in architecture.

Acknowledgements

We thank our Universidad de la Costa, Barranquilla-Colombia, for offering spaces and opportunities for professional growth and promoting academic and cientìfica production among the teaching community.

Conflict of interest

The authors declare no conflict of interest.

Author details

Carlos Augusto Rengifo Espinosa* and Oton Alberto Navas de la Cruz
Universidad de la Costa, Barranquilla, Colombia

*Address all correspondence to: creginfo@cuc.edu.co

References

[1] Demir K, Ergen B, Ergen Z, Çabuk S. Collaborative 3D design with BDMUD method: The effects of building on cityscape reflections on urban planning. ACE: Architecture, City and Environment. 2016;**11**(32):61-80

[2] Vasto D, Paola M. Influencia de las tecnologías de información y comunicación (TIC) en el proceso enseñanza-aprendizaje: una mejora de las competencias digitales. Revista Científica General José María Córdova. 2015;**13**(16):121-132

[3] Komarzyska-Swiesciak E, Adams B, Thomas L. Transition from physical design studio to emergency virtual design studio. Available teaching and learning methods and tools—A case study. Buildings. 2021;**11**(312):1-20

[4] Ching FD. Arquitectura. Forma, espacio y orden. Mexico: Editorial Gustavo Gili; 2015

[5] Acampa G, Crepo-Cabillo I, Marino G. Representación del dibujo frente a simulación de los sistemas BIM. Oportunidad o amenaza para la arquitectura. ACE: Architecture, City and Environment. 2019;**14**(40):111-132

[6] Gómez-Tone H, Martin-Gutierrez J, Bustamante-Escapa J. Spatial skills and perceptions of space: Representing 2D drawings as 3D drawings inside immersive virtual reality. Applied Sciences. 2021;**11**(1475):1-23

[7] Escorcia-Oyola LI, Jaimes de Triviño CA. Tendencias de uso de las TIC en el contexto escolar a partir de las experiencias de los docentes. Educación Y Educadores. 2015;**18**(1):137-152

[8] Arango Lozano C. Centennials: Generación sin etiquetas. Bogotá: Fundación Universidad de Bogotá Jorge Tadeo Lozano; 2019

[9] Sacker García J, Bernal Martínez MP. Pedagogía desarrollista en la práctica del docente de Ciencias Económicas de la Universidad de la Costa. ECONÓMICAS CUC. 2013;**34**(1):55-84

[10] Celani G. Enseñando diseño generativo: una experiencia didáctica. Sao Paulo, Brazil: Universidad Estatal de Campinas; 2009. pp. 162-165

[11] Oktay HE, Mutlu DH, Unvan M, Kavas KR, Bakır İ. Virtual education trials and evaluation process in architecture. Journal of Qualitative Research in Education. 2021;**25**:302-315

[12] Wu W-L, Hsu Y, Yang Q-F, Chen J-J. A spherical video-based immersive virtual reality learning system to support landscape architecture students' learning performance during the COVID-19 era. Land. 2021;**10**:561

[13] Herrera J. La investigación cualitativa. Recuperado en 2021, a partir de. 2017. Available from: https://juanherrera.files.wordpress.com/2008/05/investigacion-cualitativa.pdf

[14] Monje Álvarez C. Metodología de la investigación cuantitativa y cualitativa Guía didáctica. Neiva. Recuperado a partir de. 2011. Available from: https://carmonje.wikispaces.com/file/view/Monje+Carlos+Arturo+-+Gu%C3%ADa+did%C3%A1ctica+Metodolog%C3%ADa+de+la+investigaci%C3%B3n.pdf

[15] Anguera Argilaga MT. La investigación cualitativa. Educar. 1986;**10**:50

[16] Goñi Vindas A. Desarrollo de la creatividad. Costa Rica: Editorial Universidad Estatal a Distancia; 1999

Chapter 6

A Participatory Content Authoring Workflow for Augmented Reality at Industrial Maintenance

Camila Rossi, Marinilda Lima, Alex Álisson Santos and Ingrid Winkler

Abstract

The adoption of augmented reality-based instructions enhances maintenance operations by shortening job completion time and reducing errors. However, scaling augmented reality in industrial settings remains costly since content authoring demands computational skills such as 3D modeling and programming. Furthermore, processes can easily become obsolete, causing maintainers to abandon written instructions. So, we propose an augmented reality-based participatory content authoring workflow for maintenance tasks. We followed the Design Science Research paradigm, which included a literature review, the conception of a workflow, and a simulation to evaluate the proposed workflow's validity. We found that current workflows overlook participatory content authoring involving maintainers and that most research focuses on describing the technical architecture of proposed systems rather than a workflow that supports the use of technology in industrial settings. Regarding our proposed participatory workflow, most respondents stated it was simple to use, improved their capacity to develop augmented reality content and would help the industry adopt augmented reality. As a result, our participatory authoring workflow can optimize augmented reality content authoring during maintenance, encouraging the maintainers' interaction, and provide opportunities for procedure improvement. We conclude that non-programmer-friendly augmented reality software tools save content production time while enhancing users' perceptions of their own technological talents.

Keywords: augmented reality, maintenance, workflow, industry 4.0, authoring tools

1. Introduction

Maintenance is critical in industrial organizations, boosting competitiveness, and contributing to long-term development [1]. Maintenance operations are knowledge-intensive for maintainers because they frequently involve massive amounts of equipment, subsystems, components, and high-complexity tasks spanning from machine diagnostics to repair [2].

Although valuable, knowledge acquired in an organization can soon become obsolete, whether owing to the replacement of knowledgeable professionals with less experienced personnel, poorly standardized activities, technology obsolescence, insufficient record keeping, and loss of processes [3]. Furthermore, rising industry competitiveness and fast technology advancements have resulted in the requirement for each employee to operate more efficiently and effectively, making information readily communicated between experts and novices [4]. To achieve these requirements, knowledge management plays a vital role in establishing systematic methods of identifying, acquiring, and deploying an organization's intellectual capital [5]. As a result, the knowledge transfer process, described as "the transportation of knowledge from one place, person, system, or property to another" [6], is critical for the maintenance's long-term viability.

Companies that produce and disseminate knowledge throughout the organization, integrating it into new technologies and products, are successful [7]. This knowledge might be tacit (learned by experience over time, very personal, and difficult to communicate) or explicit (expressed in words, and shared in the form of data, visual, sound, product specifications, or manuals). When tacit knowledge becomes explicit knowledge, a transfer occurs an "externalization" [7]. In maintenance, this is frequently accomplished through the development of processes.

A maintenance process is a series of operations carried out by one or more specialists in order to keep an item's dependability. The method comprises descriptive text and may incorporate diagrams and illustrations to help maintainers comprehend the material. Although the instructions were originally printed, it is now usual to obtain them electronically [8].

The adoption of augmented reality (AR) procedures in maintenance has been found to save time and reduce mistake rates. AR technology supplements the real world with virtual objects (generated by a computer) that appear to coexist in the same space as the real world [9]. Nevertheless, the application and scalability of AR in an industrial environment continue to have a high development cost, particularly in its content authoring process, because numerous computational skills such as modeling, programming, and animation are still required.

Furthermore, present procedures (electronic or printed) are frequently perceived as irrelevant or outdated by maintainers since they do not reflect the reality of the operation [10]. Change is a constant in industrial processes, forcing the ongoing modification or updating of one or more procedures [11]. Most of the time, these procedures are written or revised by specialists with limited actual operating expertise, resulting in a gap between knowledge and industrial reality [12]. Additionally, procedures are rarely required to accomplish an activity [11]. Most of the time, maintainers know what to do and can adapt to changing conditions. As a result, the efforts necessary to create procedures are often given low priority, and their deployment is delayed. Even when procedures are accurately defined, updating them is often given low priority, resulting in obsolete knowledge and a loss of confidence among operators and maintainers.

As an example, in a study with 400 maintenance technicians and managers from industry, the interviewees were asked to propose improvements to contribute to greater use of the procedures, and the most frequent suggestion was "to involve end users in the development" [12], corroborating that the participation of the maintainers is critical for the successful adoption of the procedures by its users [11].

Thus, while conventional generating maintenance procedures generally do not involve user participation and result in outdated and irrelevant processes that are poorly accepted by users, augmented reality may improve maintenance procedure efficiency, but AR-based content development still necessitates computational abilities such as modeling, programming, and animation [13].

In this context, we propose a participative content authoring workflow for maintenance tasks based on augmented reality.

This paper is organized as follows: Section 2 describes the materials and methods used, Section 3 describes our findings, and Section 4 presents our conclusions and suggestions for further investigation.

2. Materials and method

We adopted the Design Science Research (DSR) paradigm. In addition to a knowledge contribution, DSR contributes to the real-world application environment from which the research problem or opportunity is drawn [14].

Our method parallels that described by Gregor and Hevner [14], which includes six steps: (1) identify the problem; (2) define solution objectives; (3) design and development; (4) demonstration; (5) evaluation; and (6) communication.

In steps 1 and 2, we realized the problem situation and defined the solution objectives, which is to develop an augmented reality-based participatory content authoring workflow for maintenance tasks.

In step 3, a systematic literature review was designed. Initially, we conducted a preliminary examination of 10 works addressing the usage of augmented reality and the efficiency of maintenance processes [15]. Then, this preliminary approach was improved using the strategy provided in Ref. [16]. The phrase "((framework OR workflow) AND (augmented reality OR mixed reality OR extended reality) AND maintenance)" was searched in the databases IEEE Xplore, ScienceDirect, and Scopus to answer two research questions: "Q1: Are there any workflows proposed for augmented reality authoring content that allow participatory content creation among end-users?" and "Q2: What are the differences and similarities between the workflows encountered?" The papers were then screened and criteria were applied to include works of type articles, magazines, and conferences published since 2013 and excluded works that were not applicable for industrial maintenance and did not provide an augmented reality workflow. We found nine papers that were categorized by the authorship type indicated by Ref. [13]. **Table 1** shows the results of this strategy.

In step 4, in light of a typical authoring workflow used in the manufacturing industry, we demonstrate a proof-of-concept of our proposed workflow of participatory content authoring in AR at maintenance, tested and revised it through expert reviews, and exposed preliminary versions to researchers in seminars and workshops, including International Symposium on Innovation and Technology (SIINTEC) and Brazilian Maintenance and Asset Management Association Conference (ABRAMAN).

In step 5, we evaluated the proposed workflow in terms of validity criteria, since DSR includes gathering evidence that the artifact is useful, meaning that the artifact works and does what it is intended to do [14]. We evaluated our workflows' validity by carrying out a simulation with 10 manufacturing specialists divided into three teams based on their similar skills. The age group of the participants was 50% between 20 and 30 years old, and 50% between 31 and 48 years old; eight participants had skill and interest in new technologies; eight participants were experienced with the use of augmented reality in maintenance; and seven participants had experience in creating procedures, despite the fact that only two participants considered themselves capable of creating AR maintenance procedures. The maintenance procedure simulation was performed on a Cessna C500 aircraft (**Figure 1**) and consisted of five stages.

Ref.	Authoring Type	Participatory Authoring?	KPI
[17]	By annotations	YES	N/A
[18]	By annotations/manual	YES	Time
[19]	By annotations	YES	N/A
[20]	By boxes	NO	Time and Error Rate
[21]	Manual	NO	N/A
[22]	By annotations/manual	NO	N/A
[23]	Automated	NO	N/A
[24]	Manual	NO	N/A
[25]	Manual	NO	N/A

Table 1.
List of retrieved works.

Figure 1.
Aircraft C500 Cessna subjected to the simulation.

At **Stage 1: ar creation and visualization tool training**, a 60-minute tutorial on the AR content authoring tool was given to the 10 participants. The Viscopic Pins software was utilized, which was designed for non-programmers to create AR procedures. The participants were shown how to produce AR content in the tool and the different forms of 2D (**Figure 2a**) and 3D (**Figure 2b**) content visualization. Following the demonstrations, participants put the authoring tool to the test and developed original content based on what they had learnt. Each team had 10 minutes to utilize the tool. A 3D CAD drawing, images, and videos were made accessible for use in the procedure's development.

In **Stage 2: aircraft inspection training**, the participants received a 30-minute seminar delivered by an aircraft maintenance expert, with the knowledge relevant to the simulation to be performed.

The Aircraft inspection simulation with AR procedure took held at **Stage 3: aircraft inspection simulation with AR procedure**, with the groups being prompted to perform three macro inspections: oil check, attack board fuselage, and supply valve.

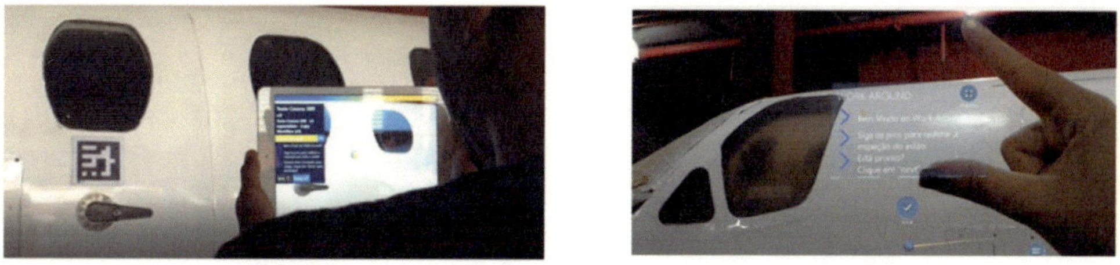

(a) (b)

Figure 2.
2D (a) and 3D (b) content visualization.

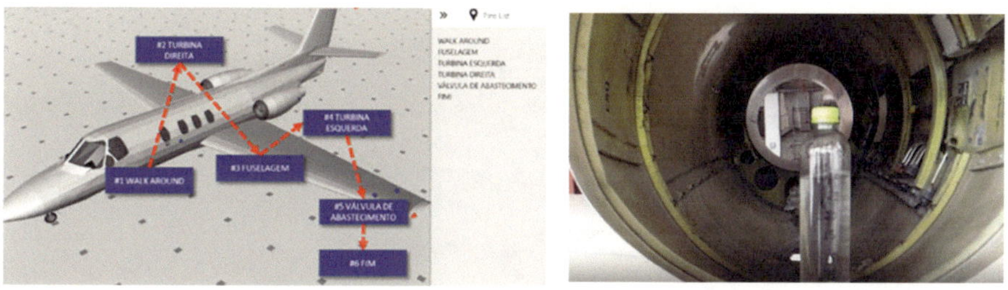

(a) (b)

Figure 3.
Maintenance simulation route (a) and oil check stage (b).

The precise location of where the inspections should be performed was included in the AR method and followed the route depicted in **Figure 3a**.

Because the stages did not follow an ordered flow, the participants suggested that the sequence of inspections be changed as a chance for improvement. In addition to completing the checklist, if any of the stages were not completed within the time frame set, the participants would be required to conduct some action to rectify the situation. During the oil check stage, the oil level in each turbine (represented by a plastic bottle of water, as shown in **Figure 3b**) had to be checked and refilled as needed. The amount of oil required was not specified in the AR method. This information was included in the materials given to the crew on a clipboard and on a page near the warehouse. The turbines required 30 liters of oil, however, the left turbine had only 15 liters. The oil container was in the warehouse, ready for the teams to fill. A picture in the attack board's fuselage revealed the exact spot (marked in red) where the crew should count the missing rivets and place them if necessary. The rivets were accessible at the warehouse, and it was essential to replace the 16 rivets that were missing. Furthermore, the complete 3D leading edge appeared overlaid in real time with AR (**Figure 4**).

Finally, at the supply valve stage (**Figure 5**), the groups were required to seal the valve cover by softly tapping it (information supplied verbally during training), in addition to locking it. In augmented reality, a snapshot of the closed valve was made available.

The groups developed a competitive environment. The procedure's time and mistakes were recorded. A minute was added to the overall inspection time for each error committed by the group. If the instructions supplied to the task competition were insufficient, the groups might contact the aircraft expert for assistance, but each assistance added 1 minute to the task completion time.

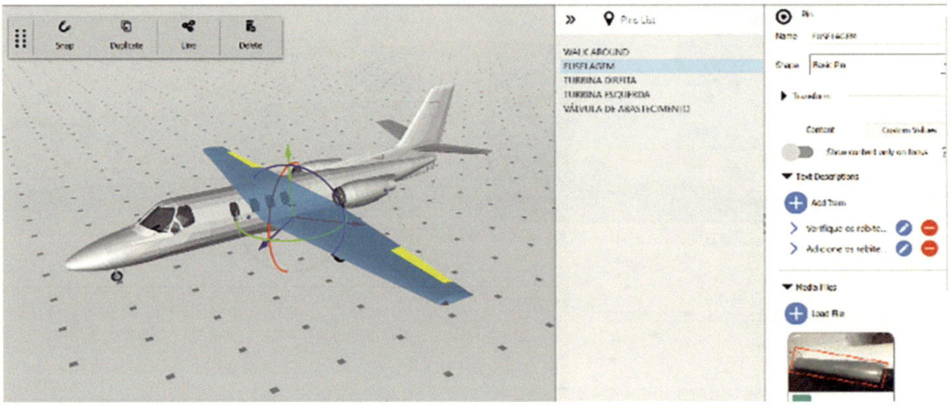

Figure 4.
AR Fuselage stage.

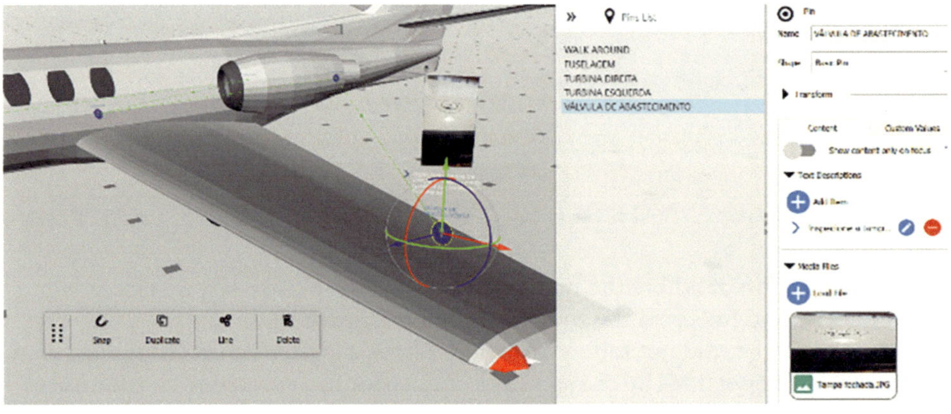

Figure 5.
AR supply valve stage.

The participants examined what information might be added to the AR procedure at **Stage 4: update of AR procedure**, guaranteeing that the instructions would be full without the usage of paper or oral directions. Each group was given 1 minute to shoot photos or videos.

At **Stage 5: aircraft inspection simulation with updated AR procedure**, the improvements recommendations were reviewed to develop an ideal procedure, and the participants engaged in the aircraft Inspection collectively to evaluate the result. Participants answered a post-dynamics questionnaire at the end of the simulation.

Step 6 of the Design Science Research approach entails communicating our findings in this work.

3. Results and discussion

In the sections that follow, we describe our findings from the literature review, the development of the proposed workflow of participatory content authoring in AR at maintenance and the simulation to evaluate it.

3.1 Participatory content authoring workflows for AR at maintenance

The systematic literature review yielded nine works, which are listed in **Table 1**. These papers were categorized based on the authoring type specified by Ref. [13], whether the addressed workflow suggested participatory creation with end-users, and whether any Key Performance Indicators (KPI) were used to demonstrate the efficacy of the workflow.

Refs. [17–19, 22] addressed workflows that allow the interaction of users through virtual annotations or block edits of the system, allowing participative content authoring.

The solution developed by [17] allowed maintenance technicians to edit the positioning and content of information in AR using a mobile application. The technician's edition was sent to the content developer to validate the information and make it available on the system, naming this type of authoring "bi-directional." Content authoring was also allowed online for technicians who were doing remote maintenance, assisted by another technician who could create the content in AR and make it available in real time to another maintainer.

The system proposed by Geun-Sik et al. [18] was developed in the aviation industry and consisted of three modules: augmented reality (AR) module; knowledge-based system module (KBS); and a unified platform with an integrated UI/UX module between AR and KBS. At the time of maintenance, the system recognized the component and made the most relevant information for that activity on the screen using artificial intelligence (AI). Despite the use of AI, the manuals in the KBS module were made manually by specialists in the field. Within the tool, it was possible to include notes from users to specialists.

Holger et al. [19] developed an AR-based approach using virtual notes as a communication platform between maintainers. These notes can be attached to components without predefined markers, eliminating the need to configure the AR system. In addition to textual information, these notes can refer to different interaction methods, such as factory process data, manuals, audio, video, and so on.

Some studies did not propose participative authoring. Lamberti et al. [20] presented the structure of the software developed in the EASE-R3 project, which explores reconfigurable AR procedures and remote assistance to overcome some of the limitations of current solutions. Michael et al. [21] proposed a system to improve communication between maintainers in the workplace. Despite providing real-time data from the machine, the system does not propose instructions on how to carry out an activity. However, it provides the status of the activity carried out by a maintainer to notify other interested parties when they can proceed with other activities.

Dimitris et al. [23] presented the development of an augmented reality application for a CNC bending machine. From the equipment's CAD information, the system was able to create the necessary instructions for configuring the machine, as well as make available the 3D step-by-step that already existed before the application. A feature was also incorporated into the system that made it possible for the maintainers to communicate with engineers, effectively addressing possible problems in production.

Siew et al. [24] proposed the ARMS system, which provides maintenance support by integrating three modules: visualization, haptic (tactile), and detection based on markers. The developed system provides feedback to users regarding the location of maintenance using tactile sensors. In addition, the content available to the user is different depending on their level of knowledge. This function is automatically changed if one of the stages to be completed by the maintainer takes longer than expected or manually by the user himself.

In the workflow proposed by Havard et al. [22], four departments were approached: design, information technology (IT), specialists, and operation. The solution mixes RA and RV in the flow, and each department has a well-defined responsibility. The design develops the 3D model of the system; TI creates the AR and VR library; the specialists develop the interactions; and users, if they are performing a task for the first time, undergo VR training first and then proceed to do the task with AR assistance. At the end of the activity, the user can give feedback to the specialists who will define the need for updates. Despite not proposing a maintenance KPI to validate the workflow, the authors reported a first analysis indicating that a developer takes 96 minutes to create an operation stage in RA without the solution. With the solution, this process decreases to 51 minutes.

Zubizarreta et al. [25] proposed a model composed of a set of software libraries and two main applications: an authoring tool that creates a database with information on the steps to be performed, including several types of multimedia information (text, video, VR and AR animations) and a guidance tool that presents that information using a mobile device. Both tools are designed to provide a simple user interface. The user can access a machine's process information and navigate through each stage of a maintenance operation. An interface also allows the creation of new content (for the authoring tool) or its visualization (using the guidance tool). Although tracking works with markers or based on the model, importing the equipment's geometry is necessary for the system to recognize the positioning of the content. The study focuses on the technical approach of image recognition, and the tests carried out measured indicators of system performance and not its impacts on users' efficiency.

The majority of addressed workflows [17, 20–25] do not consider the participative authoring of augmented reality content involving their end-users, in this context, the maintainers. Most studies are concerned with describing technical details about the architecture of the proposed systems instead of a workflow that facilitates the use of technology in an industrial environment. In addition, except for Ref. [18, 20], all the others did not use KPIs to measure the impacts of their workflows on maintenance. Among the proposals that allow the participation of end-users in the development of procedures, only Ref. [17] developed a workflow with the interactions between the stakeholders of the process (developer and user), with the rest presenting a model of the system architecture.

As a result, there is a potential to design a workflow that, in addition to allowing non-programmers to author content, involves the participation of its end users, therefore contributing to the industry's adoption of AR.

3.2 Proposition of a participatory content authoring workflow for AR at maintenance

Maintenance procedures are typically developed by a professional with minimal practical experience, resulting in a detachment from industrial realities [12]. Furthermore, the process of developing a procedure requires a significant amount of time from the specialists concerned, which might jeopardize its updates [11]. **Figure 6** depicts how procedures are frequently generated in the manufacturing industry, exemplifying the procedure creation cycle outlined in Ref. [11].

To start, the maintenance engineer is in charge of creating the digital procedure in a text editor. The Quality Engineering sector makes a template with the company's internal regulations available on the document management system (DMS). The procedure's content will be determined by the maintenance engineer's knowledge. He will be able to apply his tacit knowledge, arrange meetings with maintainers and

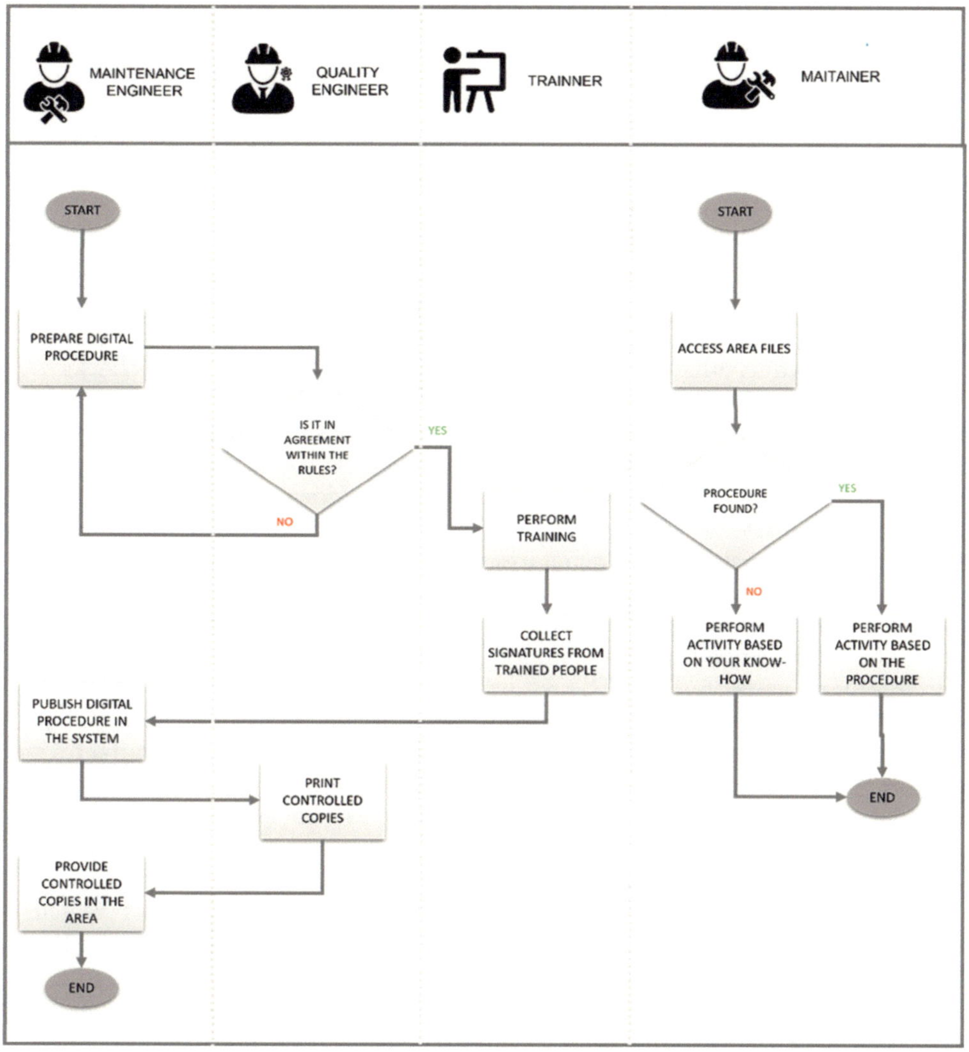

Figure 6.
A typical workflow for creating maintenance procedures in the manufacturing industry.

specialists, and review manufacturer's instructions. Photos can also be included by the maintenance engineer to help explain the methods outlined. The maintenance engineer prepares the procedure and uploads it to the DMS, where it is reviewed by the quality engineer. It is crucial to note that this procedure does not occur for all machine faults, but just for those that the maintenance engineers judge to be the most critical.

The quality engineer then verifies that the procedure is clear, adheres to safety and quality requirements, and follows the company's internal norms. If something does not meet the standard, the document is returned to the maintenance engineer until the problem is resolved. After the quality engineer has authorized the procedure, it is required to contact the training department before it is made available in the system to all employees.

Then, the trainers in each area are responsible for carrying out training for that procedure and collecting physical signatures from each trained maintainer. In many instances, the department itself creates a presentation with videos, so that the information is understood more clearly, however, this material is not available for later consultation at the DMS. In addition, there is no urgency to train maintainers, as standard operating procedures, that is, instructions for operators who work directly with the product, are prioritized to maintain the quality and safety of the process.

The maintenance engineer will then make the document available on the DMS, and the quality engineer will print physical copies.

Finally, the maintenance engineer removes the printed procedures from the quality department and places them in a physical folder at the area where the activity occurs. This stage can take a long time for the maintenance engineer depending on the number of areas that will employ these procedures, and it is distant from the reality of the maintainers, who seldom use physical manuals.

In the meanwhile, when a maintainer must do a maintenance task, he checks to see if the procedure is accessible in the physical files or the DMS. If the procedure is discovered, the activity is carried out with the assistance of the instructions created; if the procedure is not discovered, the activity is carried out based on their knowledge.

The availability of current procedures requires time due to workflow constraints. Because it is dependent on training and signature collection prior to disclosure, and because printing physical copies generates rework for engineers who have already made digital copies available, maintainers are only involved if the maintenance engineer chooses to request assistance in the creation of the instructions; otherwise, there is no direct feedback from the maintainer to the maintenance engineer.

In this work, we propose a participatory content authoring workflow for AR at maintenance (**Figure** 7), based on the workflow described in Ref. [17] and taking into account the limitations of the procedure creation pipeline in the manufacturing industry.

As part of the workflow, an authoring tool that allows any user to produce AR content without programming is required.

Our workflow takes into account just the end user (maintainer), a specialist (maintenance engineer) who will assess the approach the information was created, adding more data as needed, and an auditor (a quality engineer) to ensure that the procedure adheres to the company's rules.

The training department was eliminated since it is considered that AR technology provides the end user with the essential comprehension of the process.

The user will first open the AR application and search for the procedure to begin. There are two options from there:

- If the procedure cannot be found, the user must complete the task without the aid of AR.

- When the activity is over, the user creates an on-site procedure (with images, videos, and/or text components), saves it, and exits the app.

- If the procedure is found, the user accomplishes the activity with AR assistance. At the conclusion of the activity, the user is asked if the task was accomplished simply by following the instructions. if the response is:

  - No: An opportunity for improvement was discovered. The user must alter or add the portion in question, save, and exit the application.

  - Yes, simply exit the application.

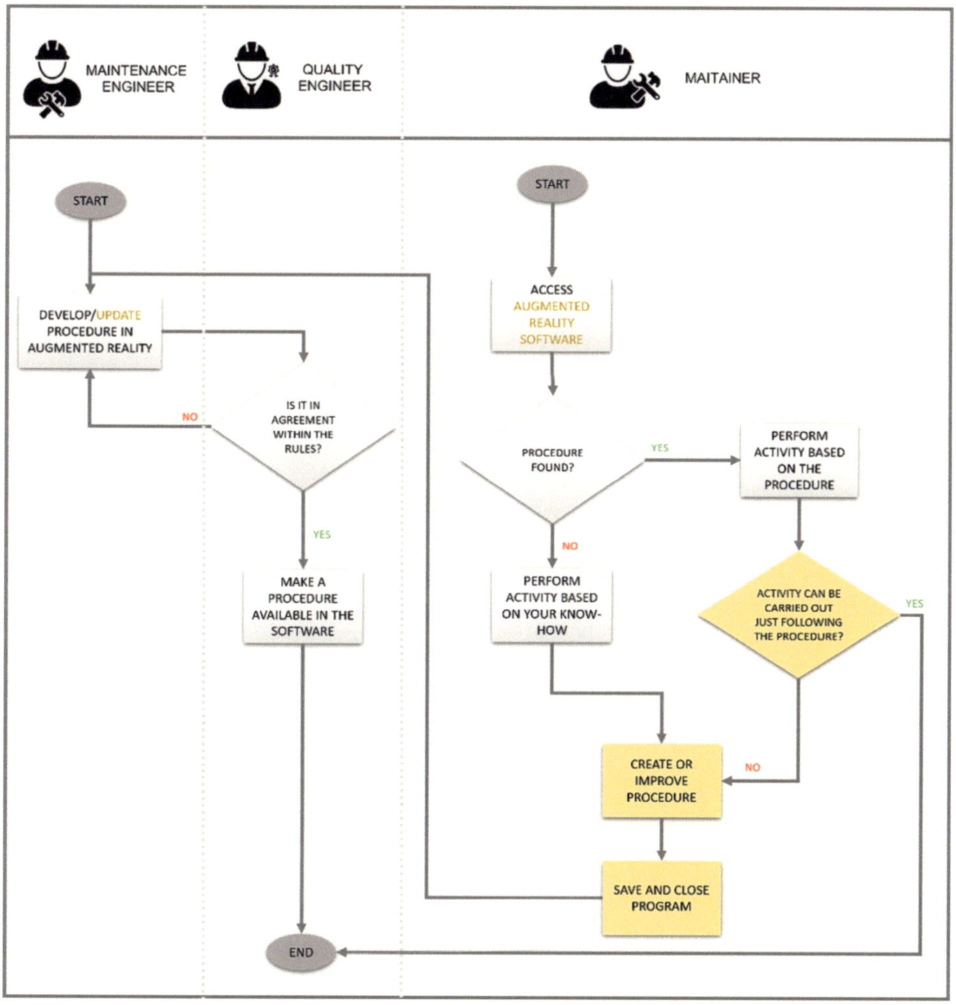

Figure 7.
Our proposed participatory content authoring workflow for AR at maintenance.

- When receiving updates or new procedures, the specialist determines if the content created requires more information, sends it to the auditor to conduct the conference in accordance with the company's requirements, and makes it available in the system, validating the creation.

3.3 Proposed workflow test simulation

The five stages of the simulation were completed by the three groups: augmented reality creation and visualization tool training, aircraft inspection training, aircraft inspection simulation with AR procedure, participant update of AR procedure, and aircraft inspection simulation with updated AR procedure.

Table 2 shows the improvement suggestions offered by each group.

In general, the user training time allowed all groups to propose improvements. Most improvements were offered by Group 3, whose participants had prior expertise with procedures and AR authoring. Their suggestions went above and beyond what was

	Group 1	Group 2	Group 3
Change stage	x		
Identify the oil level in the turbines	x	x	x
Add valve closure video	x		x
Describe Turbine Position	x	x	
Describe valve side		x	
Identify required security items			x
Check the integrity of rivets			x
Check valve integrity			x
Verify signal light			x
Request date and signature at the end of the procedure			x
Total of improvement suggestions	4	3	7

Table 2.
Suggestions for improvement from groups.

required to examine in the checklist (for example, check signaling light), as well as the inclusion of an alert on the personal protective equipment (PPE) required to carry out that inspection. We infer that because the group has more expertise in creating procedures, the improvements emerge in a more intuitive manner than the other groups.

The only group to suggest a modification in the order of stages was group 1 (**Figure 8**), which was formed by participants who had no experience in creating procedures.

As noted, the lack of order, following the inspection, was included on purpose in the initial design of the activity in order to force the teams to go further than required. In other words, this improvement idea directly contributes to reducing total inspection time.

All groups argued that the AR procedure previously informed them of the right amount of oil to be poured into each of the plane's turbines (avoiding the use of

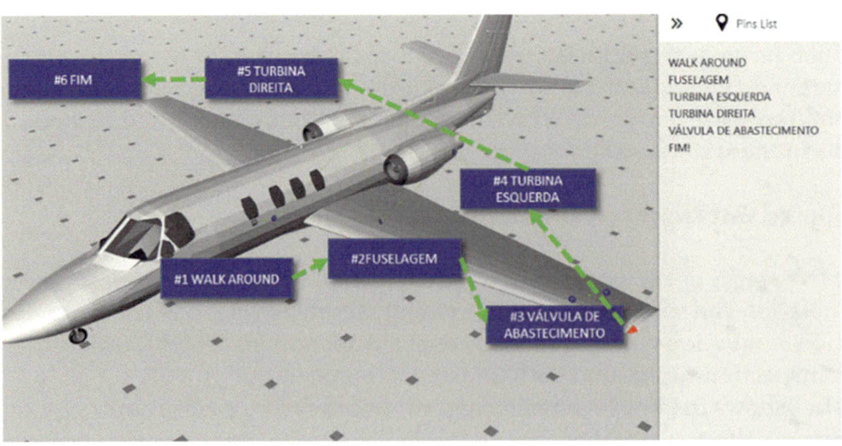

Figure 8.
New route proposed by group 1.

paper). Two groups proposed, including greater information on the plane's left and right sides to improve procedure accuracy.

Each group made at least one suggestion that the others did not, highlighting the benefit of this workflow. By enabling any user to offer improvements, the procedure may be enriched in a variety of ways, leading to continuous updating and detailing.

All groups had seen the location of pins as a source of concern during the simulation. Because all groups chose 2D visualization (which displays the information on the side of the device rather than in the actual world), the groups read the information but believed they were in the proper place rather than following the lines that went to the pin. Group 2 was the most influenced by pin location since they spent more time doing the inspection simulation. As a result, it was the only group to argue that the valve location was previously given as an instruction in the RA procedure.

In terms of errors, group 1 placed the rivets on the back of the plane, entirely overlooking the location of the repair. Groups 2 and 3 just listed the amount of rivets that were missing in the right section of the aircraft on the checklist, but they did not travel to the warehouse to make the repair. These two groups likewise failed to load the aircraft with oil and were unable to locate the necessary information on the amount required. Groups 2 and 3 complained in a post-simulation conversation that it was unclear that when a deviation was discovered, it was important to effectively repair it rather than simply writing it down in the checklist.

Although Group completed the exercise in less time, it may be assumed that if the team had really completed the necessary aircraft repairs, the time would have been greater. This illustrates that, while evaluating an AR application, it is critical to examine the measurement of errors in the procedure in addition to studying the effects of technology over the course of an activity.

The groups were permitted to try the improved instructions and were instructed to use the 3D view after updating the procedure with all of the requested improvements.

At first, the participants were upset with a lack of precise step-by-step instructions in AR, leading many users to distrust the technology's ability to aid with maintenance operations. On the other hand, because all of the recommended improvements were implemented, the procedure grew more comprehensive, surprising the participants with its ability to reduce downtime.

Following the simulation, participants were asked to complete a survey on their perceptions. The key findings were:

- The final improved procedure was chosen by nine participants as the best learning method for accomplishing the aircraft inspection. Only one participant chose the aircraft specialist's spoken instructions.

- Six participants agreed that the task could only be completed by following the AR instructions, whereas four disagreed.

- Eight participants increased their view of being capable of creating AR procedures by at least one level on the Likert Scale.

- In AR, nine participants selected films as a visualization approach.

- According to five participants, one of the most challenging aspects of completing the inspection was a lack of knowledge of AR visualization. The greatest obstacle, according to four, was a lack of technical understanding of an airplane.

- Eight participants feel that our proposed workflow will make AR deployment easier in industry.

- Five participants said the workflow was simple to follow.

The findings demonstrate that following the proposed workflow, the participants felt more capable of creating AR content; yet, the lack of expertise in visualizing the authoring tool in AR was one of the interviewees' most challenging issues. More extensive training on the instrument is required in this scenario to overcome these challenges and ensure the participants' knowledge of the procedure.

Most respondents said our proposed workflow was straightforward to implement, boosted their ability to create AR content, and would contribute to the industry's adoption of AR.

As a result, our proposed participatory authoring workflow can help to optimize the authoring of content in AR at maintenance, promote end-user participation, and offer additional chances for procedure improvement.

4. Conclusion

There has been limited investigation and development of participatory workflows that involve users in procedure creation. Users' participation boosts the acceptability of the created instructions by introducing the realism of their work in the form of procedures. End users benefit from participatory creation when incorporating AR by simplifying the procedure creation process and increasing the possibilities of this technology functioning as maintenance support in their daily operations.

In terms of contributions, most research on augmented reality in maintenance focuses on technical aspects without considering the existing maintenance environment in which this technology would be used.

This study demonstrated how, from the start of implementing these systems, user participation is relevant to ensure procedure updates and raise the likelihood of adherence to usage. Otherwise, AR procedures, like digital or physical procedures, will become outdated. Identifying these gaps before adopting new technology helps increase the likelihood that these implementations will be successful.

Further study might widen the simulation to encompass a greater number of individuals, as well as studies with statistically significant samples and industrial settings. We also propose that future studies investigate how to overcome the technological limitations of AR; analyze how automated authoring would influence the proposed workflow; and analyze users' adherence to using the procedure.

Acknowledgements

The authors acknowledge the National Council for Scientific and Technical Development (CNPq) for financial support; IW is a CNPq technological development fellow (Proc. 308783/2020-4).

Conflict of interest

The authors declare no conflict of interest.

Author details

Camila Rossi, Marinilda Lima, Alex Álisson Santos and Ingrid Winkler*
SENAI CIMATEC University Center, Salvador, Brazil

*Address all correspondence to: ingrid.winkler@doc.senaicimatec.edu.br

References

[1] Starr A, Al-Najjar B, Holmberg K, Jantunen E, Bellew J, Albarbar A. Maintenance today and future trends. In: Holmberg K, Adgar A, Arnaiz A, Jantunen E, Mascolo J, Mekid S, editors. E-maintenance. 1st ed. London: Springer; 2010. pp. 5-37

[2] Roy R et al. Continuous maintenance and the future–Foundations and technological challenges. CIRP Annals. 2016;**65**(2):667-688

[3] Argote L, Beckman SL, Epple D. The persistence and transfer of learning in industrial settings. Management Science. 1990;**36**(2):140-154

[4] He F, Khim OS, Nee AYC. A mobile solution for augmenting a manufacturing environment with user-generated annotations. Information. 2019;**10**(2):60

[5] Litvaj I, Stancekova D. Decision-making, and their relation to the knowledge management, use of knowledge management in decision-making. Procedia Economics and Finance. 2015;**23**:467-472

[6] Liyanage C et al. Knowledge communication and translation–A knowledge transfer model. Journal of Knowledge Management. 2009;**13**(3):118-131

[7] Nonaka I, Takeuchi H. Teoria da criação do conhecimento organizacional. In: Takeuchi H, Nonaka I, editors. Gestão do Conhecimento. Bookman: Porto Alegre; 2008. pp. 54-90. p. 320

[8] Badler NI, Erignac CA, Liu Y. Virtual humans for validating maintenance procedures. Center for Human Modeling and Simulation. July 2002;**45**(7):56-63

[9] Azuma R et al. Recent advances in augmented reality. IEEE Computer Graphics and Applications. 2001;**21**(6): 34-47

[10] Embrey D. Intelligent procedures design: A new approach to developing operating procedures to minimize non-compliance and maximize effectiveness. In: AIChE Spring National Meeting. 2012

[11] Sutton I. Plant Design and Operations. Oxford, United Kingdom: Gulf Professional Publishing; 2017

[12] Embrey D. Preventing human error: Developing a best practice safety culture. In: Paper to the Berkeley Conference International conference Achieving a step change in safety performance. Barbican Centre, London. (EMBREY, 1999). February 1999

[13] Palmarini R et al. A systematic review of augmented reality applications in maintenance. Robotics and Computer-Integrated Manufacturing. 2018;**49**:215-228

[14] Gregor S, Hevner AR. Positioning and presenting design science research for maximum impact. MIS Quarterly. June 2013;**37**(2):337-355

[15] Rossi Camila et al. Realidade aumentada e eficiência na manutenção industrial: Uma revisão sistemática. In: Winkler I, Guarieiro LLN, Barbosa JDV, Santos AÁB, dos Anjos JP, dos Santos Amparo KK, editors. Tecnologia e Inovação: Desafio para um Mundo Global. 1st ed. Ponta Grossa, PR: Atena Editora. 2019. pp. 39-46. Available from: https://www.doity.com.br/anais/siintec2018/trabalho/63872. Available in: 20/12/2022 at 12:32

[16] Freitas FVd, Gomes MVM, Winkler I. Benefits and challenges of

virtual-reality-based industrial usability testing and design reviews: A patents landscape and literature review. Applied Sciences. 2022;**12**:1755. DOI: 10.3390/app12031755

[17] Ong SK, Zhu J. A novel maintenance system for equipment serviceability improvement. CIRP Annals. 2013;**62**(1):39-42

[18] Jo G-S, et al. An unified framework for augmented reality and knowledge-based systems in maintaining aircraft. In: Twenty-sixth IAAI Conference. 2014. pp. 2990-2997

[19] Flatt H, et al. A context-aware assistance system for maintenance applications in smart factories based on augmented reality and indoor localization. In: 20th Conference on Emerging Technologies & Factory Automation (ETFA). IEEE. 2015. pp. 1-4

[20] Lamberti F et al. Challenges, opportunities, and future trends of emerging techniques for augmented reality-based maintenance. IEEE Transactions on Emerging Topics in Computing. 2015;**2**(4):411-421

[21] Abramovici M et al. Context-aware maintenance support for augmented reality assistance and synchronous multi-user collaboration. Procedia CIRP. 2017;**59**:18-22

[22] Havard V, Jeanne B, Savatier X, Baudry D. Inoovas - Industrial ontology for operation in virtual and augmented scene: The architecture, 2017 4th International Conference on Control, Decision and Information Technologies (CoDIT), Barcelona, Spain, 2017. pp. 300-305. DOI: 10.1109/CoDIT.2017.8102608

[23] Mourtzis D et al. Augmented Reality based Visualization of CAM Instructions towards Industry 4.0 paradigm: A CNC Bending Machine case study. Procedia CIRP. 2018;**70**:368-373

[24] Siew CY, Ong SK, Nee AYC. A practical augmented reality-assisted maintenance system framework for adaptive user support. Robotics and Computer-Integrated Manufacturing. 2019;**59**:115-129

[25] Zubizarreta J, Aguinaga I, Amundarain A. A framework for augmented reality guidance in industry. The International Journal of Advanced Manufacturing Technology. 2019;**102**(9-12):4095-4108

Chapter 7

Improving Medical Simulation Using Virtual Reality Augmented by Haptic Proxy

Pierre Boulanger, Thea Wang and Mahdi Rahmani Hanzaki

Abstract

This chapter explores how the realism of haptic perception in virtual reality can be significantly enhanced with the help of the concept of haptic proxy. In haptic proxy, the position and orientation of physical objects are tracked in real-time and registered to their virtual counterparts. A compelling sense of tactile immersion can be achieved if the tracked objects have similar tactile properties to their virtual counterpart. A haptic proxy prototype was developed, and a pilot study was conducted to determine if the haptic proxy system is more credible than standard virtual reality. To test our prototype, we performed simple medical tasks such as moving a patient's arm and aiming a syringe to specific locations. Our results suggest that simulation using a haptic proxy system is more believable and user-friendly and can be extended to developing new generations of open surgery simulators.

Keywords: proxy haptic, medical trainer, virtual reality, real-time tracking, usability study

1. Introduction

In the last few years, a new generation of low-cost virtual reality (VR) systems have emerged where users can explore or interact with an artificial world in real time. Modern virtual reality systems can create multi-sensory outputs (vision, haptic, and sound), providing the illusion of real-world perception. Commercial VR systems use head-mounted displays (HMD) to monitor head movement and update in real-time the visual display mounted inside the HMD. By wearing an HMD, users can look and move around an environment and, in some cases, interact with objects with the help of two virtual reality wands (one for each hand). Developers can create these immersive worlds using game engines like Unity3D (https://unity.com/) and Unreal Engine (https://www.unrealengine.com/). Current VR technologies are used in many applications such as training, medicine, 3D cinema, video games, etc. This chapter will focus on using VR augmented by haptic proxy to develop an open surgical training system.

1.1 VR for surgical training

Efficient and safe surgical training has always been essential to the surgeon's qualification. To become a general surgeon in Canada, students must complete an undergraduate and graduate medical program, followed by a five-year residency program in general surgery. The practical component of the residency program is to observe, assist, and perform surgery in an operating room (OR) under the supervision of a fully trained surgeon. This process is slow and expensive, so developing virtual reality-based surgical simulators to prepare first-time surgeons better is paramount.

According to St John *et al.*, [1] from 2003 to 2018, the use of Minimally Invasive Surgeries (MIS) has increased in many procedures, slowly replacing open surgeries hence the proliferation of commercial VR simulators. In this approach, users are expected to grasp MIS surgical tools attached to robotic arms (typically 2) where forces are applied to the user's hands depending on the simulated interaction with the virtual patient.

Although laparoscopic techniques are gaining popularity, most procedures still use open surgery techniques. Unfortunately, open surgical simulators are difficult to develop using current haptic technologies as they do not provide sufficient perceptual feedback to be persuasive. This is troublesome since haptic feedback is one of the primary senses surgeons use to perform these tasks. Good haptics rendering for open surgery should help to perceive the surface texture, elasticity, and temperature of objects through the mechanoreceptors in both hands. It should also help detect and manipulate objects without or with reduced visibility (e.g., in the dark).

In this chapter, we will explore how the concept of haptic proxy can be used to help develop a new generation of open surgical simulators.

1.2 Chapter organization

Section 2 will look at advanced medical simulation using VR and haptic. Section 3 describes the proposed enhanced VR system using a haptic proxy and the baseline VR system using no haptic returns. Section 4 describes experimental protocols comparing standard VR with VR augmented with haptic proxy for three simple medical procedures. A statistical analysis of each procedure is presented. We then conclude with Section 5, looking at the potential of the haptic proxy approach to creating a new generation of open surgical simulators.

2. Related work

Many researchers have attempted to develop new ways to make the virtual reality experience more realistic, particularly for surgical training. In conventional VR devices, most systems concentrate on the visual and auditory senses (i.e., they are our dominant senses). Many commercial systems such as Oculus (https:/www.oculus.com) and HTC Vive (https://www.vive.com) have been developed based on those two senses because it is easy to create realistic experiences. Very few commercial systems have used the sense of touch (haptic feedback) to complement vision and audition. Because the sense of touch is crucial to surgical training, three approaches can be found in the literature. They comprise simulators with haptic robots, haptic exoskeletal gloves, and haptic proxies.

2.1 Simulators using haptic robot

For reasons discussed previously, research and development on surgical training systems using virtual reality simulators have focused primarily on laparoscopic surgeries. Laparoscopic surgery is well suited to the current capabilities of haptic interface technology because the tools can be easily attached to force feedback systems. Current systems are realistic enough to allow for a direct transfer of competencies to the OR. Training in an immersive environment has proven that it can shorten the learning time and reduce adaption time for military and ICU staff [2, 3].

2.1.1 Simulators using haptic robot and 2D display

Let us first reviewed three of the most popular commercial virtual reality training systems for laparoscopic surgery, they are: MIST-VR [4, 5], Haptica ProMIS [6], and LAPSIM [7].

MIST-VR: In the research proposed by Seymour [5], the MIST-VR system included two laparoscopic tools mounted in a frame with six degrees of freedom force feedback, a foot pedal to activate simulated instruments and an accurately scaled 2D display. Research has shown that for learning gall bladder dissection, the learning rate was 29% faster for students trained in MIST-VR than for students who received standard training procedures. In addition, students trained with MIST-VR have made fewer mistakes in injuring the gallbladder or burning non-target tissues.

Haptica ProMIS: Van Sickle [6] did a study to evaluate the training performance of a complex laparoscopic suturing task by using ProMIS augmented-reality simulator. The study carried out by his team aimed to determine whether the use of the ProMIS simulator could distinguish performance between experts and beginners. The system was composed of a torso-shaped mannequin containing three-camera tracking systems to identify the instruments inside the mannequin from different angles. Furthermore, the primary camera was moved to the mannequin's pubic symphysis to generate the view on the computer screen. Experimental findings indicate that the ProMIS simulator possesses good psychometric properties. The experts did much better than the novices and had more coherent performances using the ProMIS simulator.

LAPSIM: The LapSim system is like MIST-VR, focusing on laparoscopic skills for suturing. With different modules, it simulates various tasks in the abdominal cavity [8]. The work by Duffy [7] was similar to the one by Van Sickle's research on the ProMIS simulator. The goal was to determine if the LapSim system could distinguish a beginner from an expert. The LapSim system employed a simulated laparoscopic instrument operated by a five-degree force feedback system. Test tasks included laparoscopic camera navigation, tool navigation, coordination, gripping, lifting, gripping, cutting, staple application and suturing. Compared with Van Sickles' experimental system, this system provided a more vivid and realistic simulation despite the limited degree of freedom of the haptic feedback. Duffy's study showed that the LapSim system could distinguish between the laparoscopic competencies of experts and beginners. However, the capability to transfer competencies to the operating room has not yet been demonstrated.

2.1.2 Simulators using haptic robot and HMD

Combining traditional surgical simulators with HMD makes it possible to create a highly immersive simulation environment. Huber [9] has developed the first clinical and technical feasibility study using this new configuration of laparoscopy in immersive VR. In the study, LabSim was used as a laparoscopic virtual reality simulator for "peg transfer," "fine dissection," and "cholecystectomy" tasks combining navigation maneuvers, fine preparation, and procedural aspects. Furthermore, the system used an OR model to situate the procedure in context.

More recently, Pulijala [10, 11] developed a system based on Oculus Rift and the Leap Motion devices https://www.ultraleap.com/ for training in maxillofacial surgeries. The teaching environment of Osteotomy LeFort I was used to assessing the impact of immersive virtual reality training on the confidence and knowledge of surgical residents. The actual OR environment of the system gives the simulation a sense of presence [12] and allows trainees to have a three-dimensional interaction with virtual objects. Another recent work by Ganni [13] on minimally invasive surgery training using virtual reality also proved that the adaption period was shortened with training in an immersive environment, and junior medical residents were better prepared to work in the OR.

2.2 Simulators using haptic exoskeleton gloves

A wearable haptic device can give users more precise feedback on palms and fingers and can be used to develop open surgical simulators. In theory, haptic gloves should provide users with haptic experiences through kinesthetic and tactile feedback [14]. However, subtle tactile feedback is still more challenging to achieve with the current technology. Therefore, making a system that will allow the user to "feel" the subtle texture details on an object's surface is not available. A recent review of exoskeleton actuation technology can be found in Tiboni *et al.* [15].

Pacchierotti [16] conducted a survey focusing on the fingertip and hand of the wearable haptic device. Their study provided a good definition of wearability by their form factor, weight, shape, area of interest, and ergonomics. The ability to feel softness and hardness is essential for open surgery simulations and can be simulated by simply adopting cutaneous information [17]. The friction on the finger pads is very complicated. The friction demonstrates different properties when the finger pads are dry and when the finger pads are wet (sweats) [18]. When holding a surgical tool during surgery, subtle pressure changes can be felt on the hand's fingertips. A small pressure sensor may be installed on the haptic glove to detect those changes on the finger or hand to record the indentation by simultaneously detecting the lateral and normal forces.

Based on this specification, the glove by HaptX (https://haptx.com/) using microfluidic skin can provide excellent contact haptic, force feedback exoskeleton to control fingers motion, and magnetic motion tracking to track fingers orientations. The tactile actuators displace the skin up to 2 mm to apply physical pressure that can replicate the sensation of real-world objects. Recently Keef *et al.* [19] developed a haptic glove capable of producing sensations reminiscent of the three types of near-surface properties: hardness, temperature, and roughness. Three haptic actuators are combined to accomplish this mixed mode of stimulation: vibrotactile motors, thermoelectric devices, and electro-tactile electrodes made from a stretchable conductive polymer synthesized in their laboratory. This polymer consists of a stretchable polyanion that serves as a polymerization scaffold. Participants trained to associate these sensations with roughness, hardness, and temperature have an overall accuracy of 98%, whereas untrained participants have an accuracy of 85%. Sensations can similarly be conveyed using a robotic hand equipped with sensors for pressure and temperature.

2.3 Simulators using haptic proxy

Another way to provide haptic feedback is to use a haptic proxy. The work by Oskouie *et al.* [20] uses a haptic proxy for simple tasks like typing on a virtual keyboard, as illustrated in **Figure 1**. This study compared two different ways of typing on a virtual keyboard. One uses a VR controller to select letters on a keyboard, while others touch a wall-mounted sponge, acting like a haptic proxy of the keyboard keys. Their work shows that users made fewer mistakes using the haptic proxy approach.

The work of Simeone *et al.* [21] is another example of using haptic proxy. In this research, they created two virtual worlds, a medieval piece and a spacecraft. They tried to match objects in the virtual world to objects in the physical world with some degree of accuracy. Their results showed that the participants did not report a mismatch if the virtual object they interacted with were close to the physical object. For example, if they were mostly going to interact with a handle on a box, if the handle in the physical world was close to the handle in the virtual world, they did not report a mismatch between those objects, even if those objects were not that similar.

Another paper published by Henderson *et al.* [22] uses the concept of proxy haptic in an AR environment in which they use objects available in the environment to provide haptic feedback for interacting with the system. For instance, they used the collar of a coil connector to change the value of a virtual text box. Their results showed that users could carry out the tasks of their experiment more quickly than the reference technique, virtual buttons and user interface elements projected on an undifferentiated surface.

3. Comparing standard VR to VR augmented by haptic proxy

To test our assumptions that VR augmented with haptic proxy works better than standard VR, we have developed two systems: one completely virtual called *Standard VR* and another one using the concept of haptic proxy called *Augmented VR with Haptic Proxy*. To compare both systems, we devised three simple experiments typically performed in OR; 1) Moving a patient's hand to specific locations on the chest 2) Performing surgical tool manipulations with the right arm fixed 3) Performing ambidextrous manipulations by grabbing the patient's right arm and applying a syringe to specific locations. For each of these tasks, we tested the following hypothesis:

Figure 1.
An example of a prototype virtual keyboard system using a foam board as a haptic proxy [20].

1. **H1:** The time to execute each task is longer using the system with haptic proxy than with VR controllers because of physical constraints.

2. **H2:** The haptic proxy system is more believable in terms of being closer to reality.

3.1 VR simulation system using haptic proxy

As illustrated in **Figure 2**, the inputs from user interaction with the HTC Vive beacons and wands are transferred to the simulation's computer. SteamVR (https://store.steampowered.com/steamvr) captures that data. The data are then used by the Unity3D rendering engine to update the virtual patient's pose and the syringe location.

In the proposed system shown in **Figure 3**, the proxy haptic props are composed of an articulated rubber mannequin to simulate a patient and an HTC Vive wand to

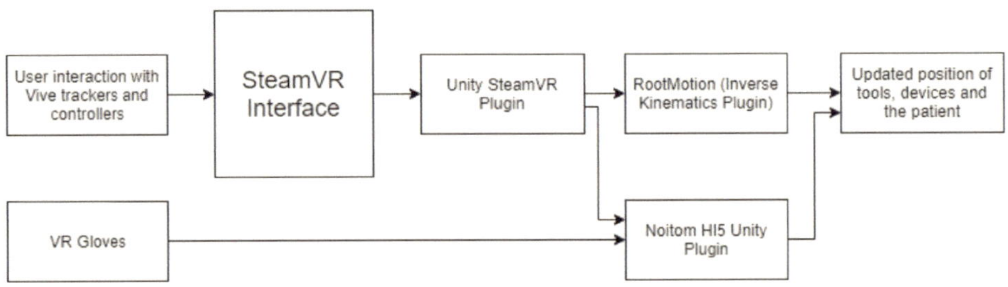

Figure 2.
Proxy haptic system architecture using Unity3D.

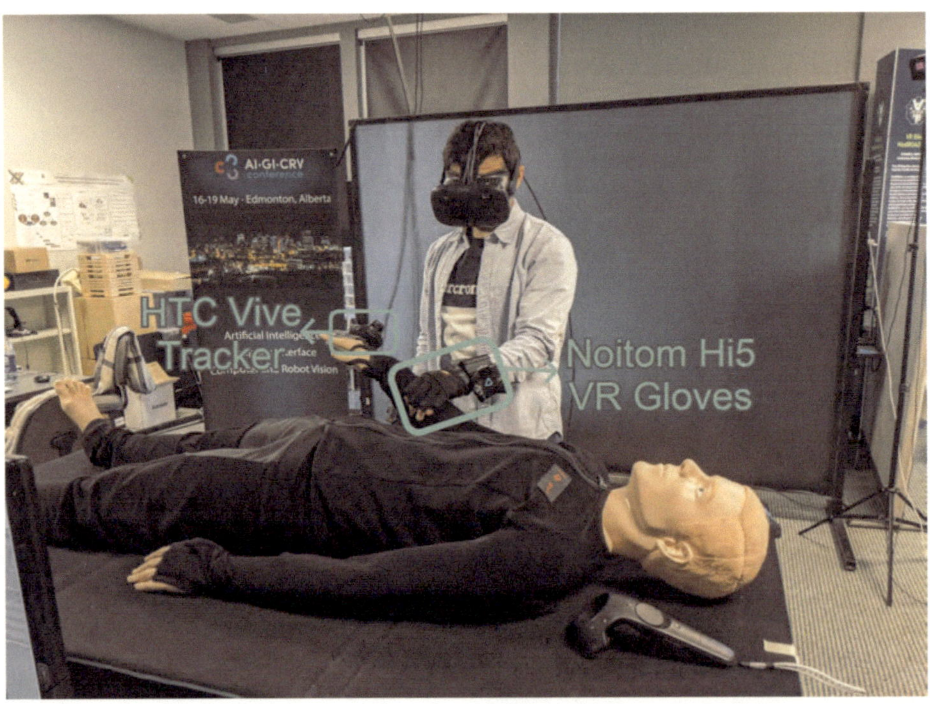

Figure 3.
Proxy haptic system composed of a tracked mannequin tracked using HTC Vive beacons and a syringe tracked by a wand.

simulate a syringe. Both props needed to be tracked and geometrically synchronized with their virtual representation. Both hand positions and finger orientations were measured using Noitom Hi5 glove. In the following section, we will describe how proxy objects are tracked and registered with their virtual counterparts.

3.1.1 Tracking and registering the proxy mannequin

The rubber mannequin representing the patient has the same weight as a middle-aged male. Although the mannequin articulations do not offer the same degrees of freedom as an actual human body, it is accurate enough for our experimentation. We explored two approaches to registering the mannequin with its virtual counterpart. The first system consisted of placing numerous retro-reflective targets on the mannequin, as illustrated in **Figure 4**. Those targets were imaged using a 12-camera OptiTrack (https://optitrack.com/) system from which a cloud of 3D coordinates was estimated using a propitiatory photogrammetric algorithm. The resulting 3D point cloud was then registered with the avatar using a skeleton matching algorithm developed by OptiTrack. A review of human joint estimation from a 3D point cloud can be found in [23].

One of the main drawbacks of this approach is its sensitivity to optical occlusions. As the user interacted over the mannequin, his body occludes critical targets

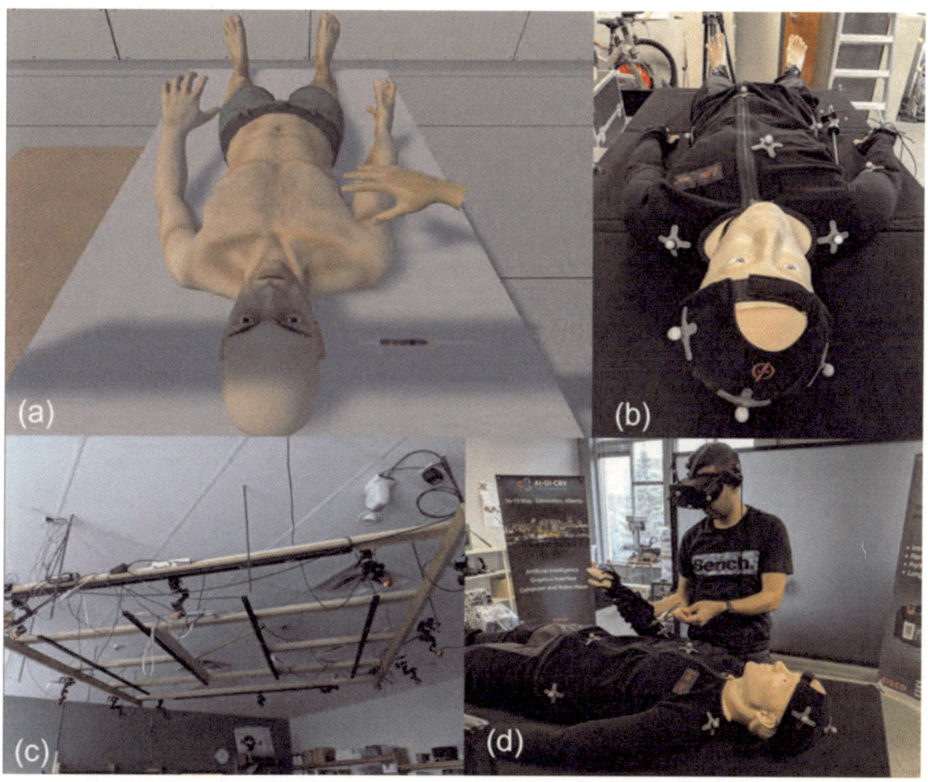

Figure 4.
Tracking mannequin motion using OptiTrack skeleton tracking system: (a) avatar representation, (b) mannequin with optical targets, (c) 12-cameras OptiTrack tracking configuration mounted on the ceiling, (d) user interacting with the mannequin.

generating faulty registrations. Because of occlusion, we concluded that this approach was impractical for our haptic proxy application, especially for open surgical procedures. One could solve the occlusion problem by using a non-optical tracking system such as Ascension 3D trackSTAR magnetic tracking system (https://tracklab.com.au/) that gives 6D (position and orientation) position of small magnetic sensors. Using this technology, each joint can be tracked in 6D and associated with the corresponding control points on the avatar. Our experiments showed that this solution worked well but is extremely costly and sensitive to the presence of metallic objects (such as surgical tools) in the environment. Finally, to animate the virtual mannequin pose, a third low-cost solution was explored. It uses an inverse kinematic (IK) approach where a limited number of HTC Vive beacons are paired with the avatar control points from which the skeleton joint configuration can be calculated directly. Using IK, one needs only to measure a few 3D points located away from occluded areas to animate joint arm configurations. For example, locating a single optical tracker on the mannequin's right hand is sufficient to animate the right arm joints. We used Unity3D RootMotion advanced character animation asset (http://root-motion.com/) to implement our approach. Since the VR model of the patient had numerous kinematic chains that needed to be animated by the skeleton using the IK engine, we needed to find the value of the ten parameters used by the IK engine that would allow the simulated motion to fit well with the real motion of the mannequin. These parameters are:

1. Hand Goal Position Weight.
2. Hand Goal Rotation Weight.
3. Shoulder Rotation Weight.
4. Shoulder Twist Weight.
5. Shoulder Yaw Offset.
6. Shoulder Pitch Offset.
7. Arm Bending Goal Weight.
8. Arm Swivel Offset.
9. Arm wrist to Palm Axis.
10. Arm Length Stretching Value.

In addition to the hand tracker, three extra optical trackers were temporally attached to the mannequin's right shoulder, elbow, and wrist to perform a one-time calibration of the IK parameters. The calibration procedure consisted of determining IK parameters that match best the four control points on the avatar model using the hand tracker as the animating control point. To calibrate, we moved the mannequin's arm to eight positions and then paired them with the mannequin VR model. The IK parameters are then computed using gradient descent methods that minimize the differences between those two sets of points. Initially, the 10 IK parameters were randomly initialized. For each set of parameters, we calculated the gradient relative

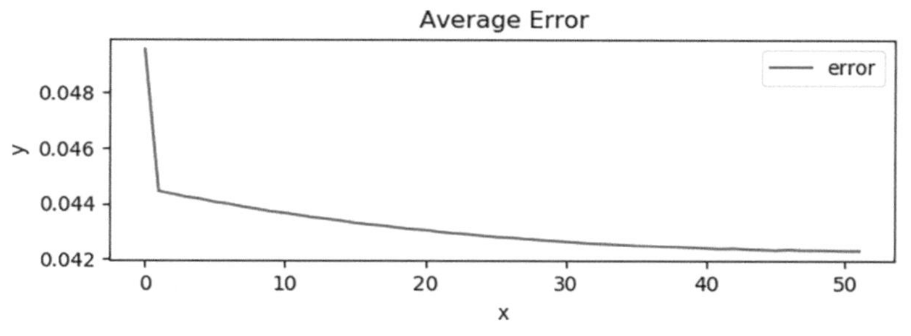

Figure 5.
IK parameters optimization: Variation of positioning error between the virtual control points and the one measured by the HTV Vive beacons vs number of iterations.

to each parameter and changed the parameters in the opposite direction of the gradient multiplied by a small gain. Using this approach, we could find a local minimum as illustrated in **Figure 5**. In the end, the IK parameters were the ones that gave us the smallest positioning error overall for the eight calibration positions. Once calibrated, only the hand tracker can be used to animate the avatar's right arm. A similar approach can be found in [24] for full body tracking using HTC Vive beacons.

3.1.2 Hand tracking

For interaction with the hand, we used the Noitom Hi5 to measure the orientation of the fingers and an HTC Vive beacon to measure the position and orientation of the wrist. One can see in **Figure 3** the system configuration. We used the available SDKs for HTC Vive and Noitom Hi5 VR gloves to integrate our system with the virtual world. With the help of the HTC Vive SDK, we could see the position and orientation of the trackers and controllers. The Noitom Hi5 VR SDK gloves enabled us to see an avatar of the hand in Unity3D along with the movement of the fingers. It also allows the user to pick up items from the virtual world.

3.1.3 Tracking the syringe proxy

For the syringe proxy, we simply use an HTC Vive wand because the base of the wand has a haptic feel similar to a large syringe. The wand tracking information is registered to a syringe model in the VR world using HTC Vive API.

3.1.4 Registration of static objects

We also had to register static proxy objects (operating table and mannequin body) with their VR equivalent. For the table, we attached two HTC Vive optical beacons at the opposite corners of the table and, using Unity3D tools, found the distance between those measured markers and the one in the virtual world. The distance between the sample markers in Unity3D was 174 cm, and the measured distance was: 176 cm. Following this measurement, we re-scaled and re-centre the VR model to match its real-world counterpart. We also registered the location and rotation of the mannequin relative to the table by installing HTC Vive beacons on the mannequin's head, chest, legs, and arms.

3.2 Standard virtual reality system

For the standard virtual reality system, we used the same environment as the system with the haptic proxy model. However, the difference was that since this was a completely virtual system, the users only interacted with the environment using HTC Vive wands. Once the users grabbed an object in the virtual world, the position of that object was updated in Unity3D based on the position of the wand if the users were holding the grab button on the controller. For the case of moving the patient's arm, we created a virtual object in Unity3D and used it as the hand end-effector, like the one used for the haptic proxy system. The users were, in reality, moving the end-effector object but what they saw was the patient's arm being moved with the wand's movement (see **Figure 6**).

4. Experimental comparison between the two systems

To test our assumptions, each task described above was carried out using the haptic proxy and standard virtual reality systems. Also, since using the right or left hand could have added a difference to the results, we decided to do each experiment with two hands. In the end, each task was done four times; **p_right** (haptic proxy with the right hand), **p_left** (haptic proxy with the left hand), **c_right** (controller with the right hand), and **c_left** (controller with the left hand). We randomized the order of these settings for different participants to reduce the learning effect on our results. Then, according to the order of each participant, we asked them to carry out the different parts of the experiments. In the end, we asked them to answer a questionnaire with some general questions on participants' information, such as their age and

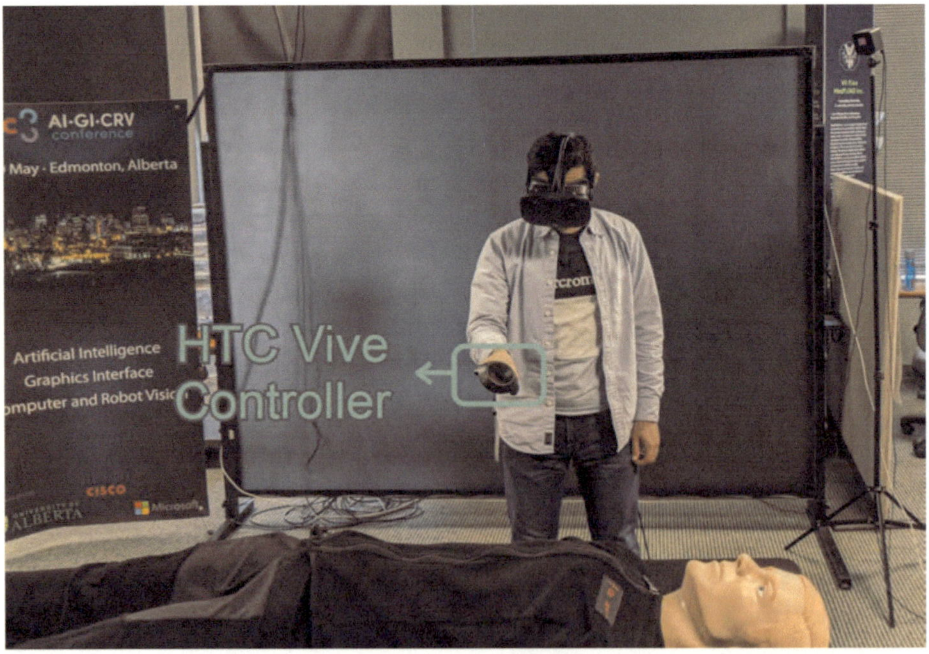

Figure 6.
Virtual system configuration: The user did not have any physical interactions with the mannequin in this scenario.

hand dominance, and some questions about their experience with different tasks using haptic proxy and controllers.

In our experiment, since it was a pilot test, we had only 12 participants (8 men and 4 women), all students at the University of Alberta, Computing Science Department aged between 23 to 33 (M = 26.6; SD = 2.6), all of whom saw in stereo. When asked about hand dominance, 11 indicated that they are right-dominant, and one was left-dominant. For the question asking "how often do you play video games" 2 of them indicated that they play video games often, 7 indicated that they play sometimes, and 3 indicated that they never play video games.

Given that we had 12 participants and each task had 10 tests, we had 120-time data for each task for each context. To find outliers in our data, we used the interquartile method [25]. In this approach, we calculate the median, 25th and 75th percentiles of the 120 data points. Next, we subtracted the 25th percentile from the 75th percentile to determine the distance between them. Then we multiplied that distance by 1.5 and added it to the 75th percentile, and we also multiplied the distance by 1.5 and subtracted it from the 25th percentile. We removed any data that was not in that range and called them outliers. We report the number of data points for each section after removing the outliers. Since we originally had 120 data points for each task, the difference in these numbers indicates the number of outliers we had for the task.

Ultimately, we analyzed the remaining data by finding the Normal and Gamma distributions with the best parameters using the Python Scipy library (https://www.scipy.org). Then we performed a Kolmogorov–Smirnov test and chose the distribution with the best p-value as the distribution on our data. Finally, we ran an unpaired t-test to compare the controller usage time with the haptic proxy system. If the best distribution found in any part of the research was Gamma distributed, then we used the mean and STD from the distribution using the Python Scipy library and performed the t-test using the means and STDs. The following sections provide a detailed description of the results of each task.

4.1 Task 1: moving the Avatar's hand to a specific location on the chest

In every trial, participants had to take place at a starting position. Then, after giving a signal from the experimenter, a timer started, a target appeared on the patient's right thigh, and participants began to move toward the patient. Next, they grabbed the patient's arm and moved it to make the right hand thumb touch the target. After that, the target disappeared, and we told them to return to the starting position for the next attempt. All times were measured in Unity3D in second. **Figure 7** shows the physical and virtual environment for VR controllers and haptics proxy systems.

4.1.1 Grabbing the arm

For experiment1a, the time to grab the arm was measured. The best distribution found for c_left, and p_right was Normally distributed with p-value 0.1325 and 0.7789, respectively. For c_right and p_left the best distribution found was Gamma distributed with p-value 0.7796 and 0.9779, respectively. The results are presented in **Table 1**.

Since the p-value for c_left is below 0.5, we only do the comparison using an unpaired t-test for the right hands. Since c_right does not have a Normal distribution we calculated the mean and STD using the Python Scipy library. The calculated mean

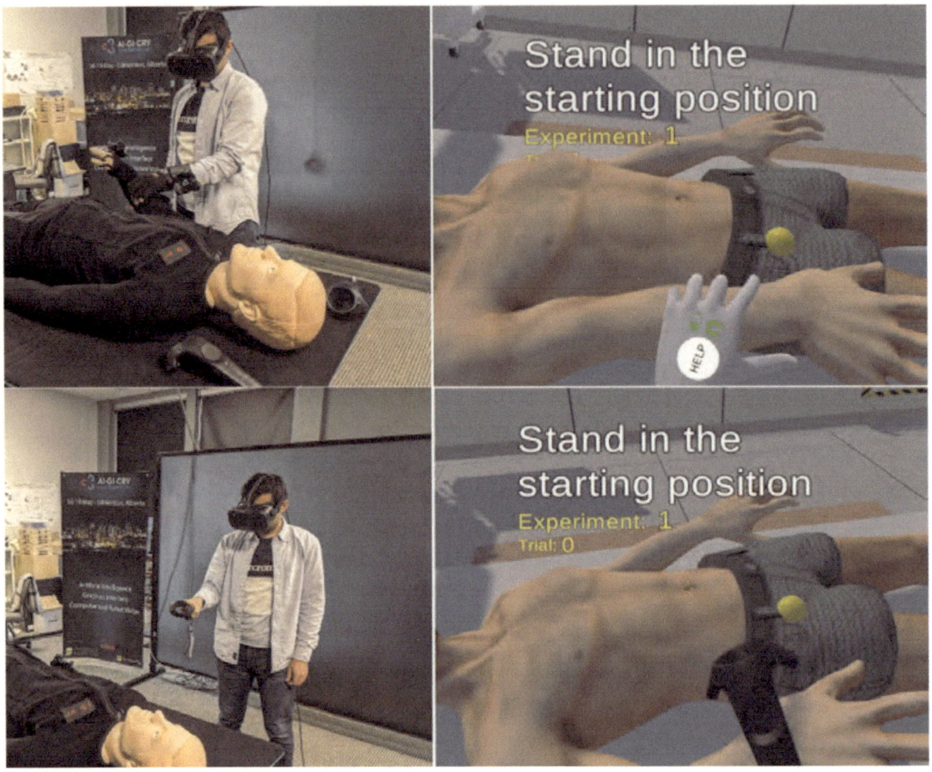

Figure 7.
Experiment 1 physical and virtual environments: In the upper left-hand corner, the participant performs the task using the haptic proxy. The upper right is what the participant sees while using the haptic proxy. In the bottom left corner, the participant completes the task with VR controllers. On the lower right, this is what the participant sees when using only the VR controllers.

Setting	Dist.		Normal			Gamma		N
		P	Mean	STD	P	α	β	
c_left	Normal	0.1325	1.2393	0.4080	0.1014	263.6384	0.0253	120
c_right	Gamma	0.5839	1.2193	0.3616	0.7796	13.0656	0.1015	120
p_left	Gamma	0.8945	1.4169	0.3651	0.9779	127.3759	0.0323	120
p_right	Normal	0.7789	1.3702	0.4119	0.3273	26.9014	0.0799	119

Table 1.
Experiment 1a: Time to grab the arm.

and STD for c_right were 1.2193 and 0.3667 respectively. After running an unpaired t-test, our p-value was 0.0031 (a = 0.05, t = 2.9919, df = 237, standard error of difference = 0.050) and therefore, based on our results, the system with haptic proxy (using the right-hand) is slower than the VR wand (controller with the right hand).

4.1.2 Moving the arm

For experiment1b the time to move the arm was measured. The best distribution found for c_left, p_left, and p_right was Gamma distributed with p-values of 0.7704,

Setting	Dist.	Normal			Gamma			N
		P	Mean	STD	P	α	β	
c_left	Gamma	0.3615	0.8217	0.2630	0.7704	3.0295	0.1611	115
c_right	Normal	0.6550	0.8002	0.2148	0.5979	2.4679	0.1488	107
p_left	Gamma	0.1374	1.4245	0.6611	0.4425	1.8480	0.5086	113
p_right	Gamma	0.1853	1.4003	0.5709	0.9563	3.4553	0.3164	109

Table 2.
Experiment 1b: Time to move the arm.

0.4425, and 0.9563 respectively. For c_right, the best distribution found was Normally distributed with a p-value of 0.6550. The results are shown in **Table 2**. Since the p-value for p_left is below 0.5 we only do the comparison using an unpaired t-test for the right hands. After performing an unpaired t-test our p-value was less than 0.0001 (a = 0.05, t = 9.9260, df = 214, standard error of difference = 0.060) and therefore, based on our results, the system with haptic proxy (using right-hand) is slower than the VR wand system (controller with right hand).

4.1.3 Returning the arm

For experiment1c the time to return the arm was measured. The best distribution found for all the settings was Gamma distributed. The results are shown in **Table 3**. In this case, the best p-value for all of them is above 0.5, we will do the t-test for both the left and the right hands. After performing an unpaired t-test for the left hand, the p-value was 0.0021 (a = 0.05, t = 3.1141, df = 231, standard error of difference = 0.028). Therefore, based on our results, in this part of the task, the system with haptic proxy (using left hand) is slower than the VR wand (controller with left hand). The p-value for the t-test for the right-hand settings was 0.1413 (a = 0.05, t = 1.4764, df = 210, standard error of difference = 0.026) and therefore our hypothesis gets rejected in this scenario.

4.2 Task 2: Surgical tool manipulations with a fixed arm location

During each test of this task, participants were asked to stay at a specific departure location. Then we gave them a signal and started the timer and the

Setting	Dist.	Normal			Gamma			N
		P	Mean	STD	P	α	β	
c_left	Gamma	0.1377	0.6502	0.1968	0.8170	5.1704	0.0877	118
c_right	Gamma	0.0314	0.6638	0.2104	0.5585	4.5355	0.0992	112
p_left	Gamma	0.1733	0.7369	0.2176	0.7451	3.1650	0.1264	115
p_right	Gamma	0.4760	0.7026	0.1650	0.5187	14.6265	0.0432	100

Table 3.
Experiment 1c: Time to return the arm.

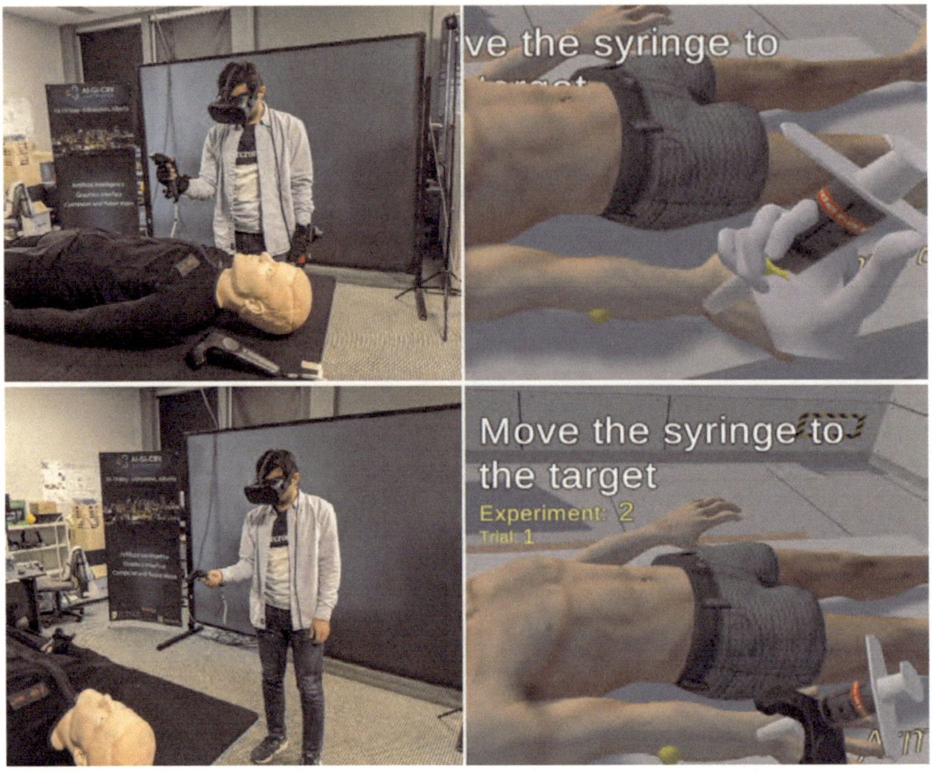

Figure 8.
Experiment physical and virtual environments: In the upper left-hand corner, the participant performs the task using the haptic proxy. The upper right is what the participant sees while using the haptic proxy. In the bottom left corner, the participant completes the task with VR controllers. On the lower right, this is what the participant sees when using only the VR controllers.

experiment. After hearing the signal, they were instructed to move to the syringe on the table and pick it up. Then they were asked to touch the tip of the syringe at some target on the patient's right arm. The target position differed for each attempt. When the middle tip came into contact with the target, the target vanished. Participants were instructed to return the needle to its original location. Participants were instructed to return to the starting position and prepare for the next trial (**Figure 8**).

4.2.1 Grabbing the syringe

For experiment2a, the time to grab the syringe was measured. The best distribution found for c_left was Normally distributed with a p-value of 0.4181. For c_right, p_left, and p_right the best distribution found was Gamma distributed with p-value 0.8982, 0.6206 and 0.9595, respectively. The results are shown in **Table 4**. Since the p-value for c_left is below 0.5, we only do the comparison using an unpaired t-test for the right hands. After running an unpaired t-test for the right-hand setting, our p-value was 0.0001 (a = 0.05, t = 3.9172, df = 237, standard error of difference = 0.050) and therefore haptic proxy system using right-hand is slower than using the controller with the right hand for this part of the task.

Setting	Dist.	Normal			Gamma			N
		P	Mean	STD	P	α	β	
c_left	Normal	0.4181	1.2258	0.4001	0.3705	1871.9184	0.0093	119
c_right	Gamma	0.5145	1.2240	0.3465	0.8982	24.7096	0.0700	119
p_left	Gamma	0.0762	1.3523	0.4289	0.6206	8.0061	0.1537	120
p_right	Gamma	0.6240	1.4208	0.4237	0.9595	29.9926	0.0775	120

Table 4.
Experiment 2a: Time to grab the syringe.

4.2.2 Moving the syringe

For experiment2b, the time to move the syringe was measured. The best distribution found for c_left was Normally distributed with a p-value of 0.8067. For c_right, p_left, and p_right the best distribution found was Gamma distributed with p-values of 0.9965, 0.9971 and 0.9114, respectively. The results are shown in **Table 5**.

After performing an unpaired t-test for the left hand, the p-value was 0.0056 (a = 0.05, t = 2.7968, df = 233, standard error of difference = 0.041). Therefore, based on our results, in this part of the task, the system with a haptic proxy (using the left hand) is slower than the VR wand (controller with the left hand). For the right-hand setting, the p-value was 0.0001 (a = 0.05, t = 4.0892, df = 231, standard error of difference = 0.046), and therefore for this part of the task haptic proxy system using the right hand is slower than the virtual system using the right hand.

4.2.3 Returning the syringe

For experiment2c, the time to return the syringe to its original position was measured. The best distribution found for all the settings was Gamma distributed. The results are shown in **Table 6**. In this case, the best p-value for c_right and p_right is lower than 0.5. Therefore, we can only do the t-test for the left hand.

After performing an unpaired t-test for the left hand, the p-value was 0.0090 (a = 0.05, t = 2.6334, df = 235, standard error of difference = 0.039). Therefore, based on our results, in this part of the task, the system with proxy haptics (using the left hand) is slower than the VR wand (controller with the left hand).

Setting	Dist.	Normal			Gamma			N
		P	Mean	STD	P	α	β	
c_left	Normal	0.8067	0.9474	0.2802	0.4893	6.1178	0.1169	118
c_right	Gamma	0.4364	0.9610	0.2913	0.9965	7.2530	0.1090	118
p_left	Gamma	0.2494	1.0616	0.3391	0.9971	4.5444	0.1608	117
p_right	Gamma	0.0861	1.1505	0.3777	0.9114	2.0299	0.2851	115

Table 5.
Experiment 2b: Time to move the syringe.

Setting	Dist.		Normal			Gamma		N
		P	Mean	STD	P	α	β	
c_left	Gamma	0.5736	0.7716	0.2658	0.8960	52.3348	0.0368	120
c_right	Gamma	0.0022	0.8241	0.2478	0.0632	5.3362	0.1042	112
p_left	Gamma	0.2952	0.8736	0.3285	0.7418	11.3005	0.0974	117
p_right	Gamma	0.0818	0.8123	0.2495	0.2683	18.1847	0.0575	106

Table 6.
Experiment 2c: Time to return the syringe.

4.3 Task 3: Ambidextrous surgical tool manipulation

In each trial, participants were expected to stand at a particular starting position. Then, upon receiving a signal from the experimenter, a timer started, a target appeared on the patient's right arm, and participants began to move toward the patient. They then grabbed the patient's arm and raised it. They were then instructed to take the syringe off the table. Next, they had to make the tip of the syringe reach the target on the patient's right arm. In this instance, they were free to move or orient the arm to touch the syringe at the target faster.

Furthermore, some targets were on the other side of the arm; that is, they had to rotate the arm to access these targets. Upon contact with the tip of the syringe with the target, the target disappeared, and they had to place the syringe back to its original position on the table, then, they were asked to put the patient's arm in its original location. Finally, we asked them to return to the starting position for the next trial. Since both hands worked on this task, there was no distinction between the right and left hands. In this scenario, what we mean by the right hand (e.g. p_right) is taking the arm with the right hand and the syringe with the left hand. Since the opposite scenario (i.e., taking the patient's hand with the left hand and taking the syringe with the right hand) required the participants to be in a cross-hand position which was difficult and not necessarily used in the real world, we only did this task with only one set. **Figure 9** shows the physical and virtual environments for both controllers-only and haptic proxy settings.

4.3.1 Grabbing the arm

For experiment3a, the time to grab the arm was measured. The best distribution found for c_right was Normally distributed with a p-value of 0.4546. For p_right the best distribution found was Gamma distributed with a p-value of 0.5591. The results are shown in **Table 7**.

Since the p-value for c_right is less than 0.5, one cannot assign a distribution to c_right, we cannot perform the t-test in this case.

4.3.2 Grabbing the syringe

For experiment3b, the time to grab the syringe was measured. The best distribution found for both settings was Gamma distributed. The results are shown in **Table 8**.

For the right-hand setting, the result of running an unpaired t-test was a p-value = 0.9129 (a = 0.05, t = 0.1095, df = 229, standard error of difference = 0.060).

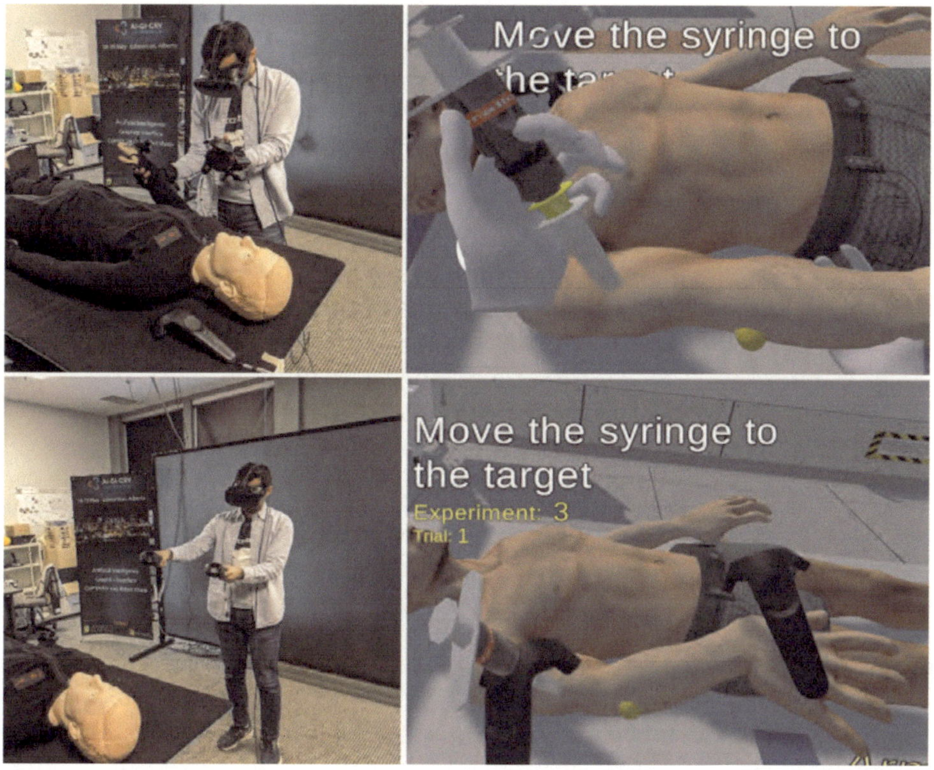

Figure 9.
Experiment 3 physical and virtual environments: In the upper left-hand corner, the participant performs the task using the haptic proxy. The upper right is what the participant sees while using the haptic proxy. In the bottom left corner, the participant completes the task with VR controllers. On the lower right, this is what the participant sees when using only the VR controllers.

Setting	Dist.	Normal			Gamma			N
		P	Mean	STD	P	α	β	
c_right	Normal	0.4546	1.1989	0.4063	0.2478	4.1734	0.2104	118
p_right	Gamma	0.4801	1.3885	0.4122	0.5591	16.4605	0.1027	118

Table 7.
Experiment 3a: Time to grab the arm.

Setting	Dist.	Normal			Gamma			N
		P	Mean	STD	P	α	β	
c_right	Gamma	0.3066	0.9155	0.3062	0.9636	6.4065	0.1225	115
p_right	Gamma	0.2467	0.9089	0.5400	0.5860	2.4630	0.3606	116

Table 8.
Experiment 3b: Time to grab the syringe.

Therefore, we cannot say anything about the controller being faster than the proxy haptics in this scenario.

4.3.3 Moving the syringe

For experiment3c, the time to move the syringe was measured. The best distribution found for all the settings was Gamma distributed. The results are shown in **Table 9**.

After performing an unpaired t-test for the right hand, our p-value was less than 0.0001 (a = 0.05, t = 4.3542, df = 218, standard error of difference = 0.041). Therefore, based on our results, in this part of the task, the system with haptic proxy (using the right hand) is slower than the VR wand (controller with the right hand).

4.3.4 Returning the syringe

For experiment3d, the time to return the syringe was measured. The best distribution found for all the settings was Gamma distributed. The results are shown in **Table 10**. Since the p-value for c_right is less than 0.5, we cannot assign a distribution to c_right, we cannot perform the t-test in this case.

4.3.5 Returning the arm

For experiment3e, the time to return the arm was measured. The best distribution found for all the settings was Normally distributed. The results are shown in **Table 11**.

Setting	Dist.	Normal			Gamma			N
		P	Mean	STD	P	α	β	
c_right	Gamma	0.5682	0.8692	0.2202	0.8399	20.6269	0.0486	108
p_right	Gamma	0.0473	1.0468	0.3568	0.8466	3.4521	0.1960	112

Table 9.
Experiment 3c: Time to move the syringe.

Setting	Dist.	Normal			Gamma			N
		P	Mean	STD	P	α	β	
c_right	Gamma	0.0120	0.7598	0.2598	0.3787	4.3089	0.1246	114
p_right	Gamma	0.0599	0.9282	0.3909	0.9879	1.9579	0.2920	117

Table 10.
Experiment 3d: Time to return the syringe.

Setting	Dist.	Normal			Gamma			N
		P	Mean	STD	P	α	β	
c_right	Gamma	0.2226	0.9712	0.3666	0.2031	29.2145	0.0682	116
p_right	Gamma	0.4484	0.9189	0.4692	0.2364	7.9018	0.1713	115

Table 11.
Experiment 3e: Time to return the arm.

Since the p-value for c_right and p_right is less than 0.5, we cannot assign a distribution to c_right, we cannot perform the t-test in this case.

4.4 Questionnaire

Our questionnaire consisted of 7 sections; Section 1 consisted of general questions about the user's physical characteristics, such as their hand dominance and if they wear glasses, and the following six questions were about the system. We had three tasks (discussed above) and two different parameters (controller or proxy haptics), and we had a questionnaire for all of them. In general, the questions for each of these six sections were identical, with minor differences depending on the section. Each section consisted of three groups of questions; **1) Perceived Competency Scale** asking questions about how confident and comfortable users were working with the system. **2) System Usability Scale** asking questions on how the system was usable in the participant's opinion **3) Believability** asking how the system was near the real world.

Each group of questions consisted of a few questions or statements about that group. The users had to choose, on a scale from 1 to 7, with seven meanings ranging from how they agreed or disagreed with each statement. Some statements were negative (for example, I found the system very cumbersome), while others were positive. In the end, we applied a function to all these responses for each group to get a score between 1 and 100. Details on scores are presented in **Table 12**. However, because they were different in nature, we could not merge all the questions in the credibility section. For example, one question asked how much they thought they were facing a real problem, and another asked whether they saw a difference between handling the right and left hands. We denoted"I felt like I am dealing with a real-world arm' as the question directly asked about the believability of the system, and we mapped the values of the responses to a number between 1 and 100 as a believability score. We will discuss other questions in the believability section later. At the end of the questionnaire, participants were also asked for their views on the system.

We next preformed an unpaired t-test with a = 0.075 for all the experiments to compare the system with only controller with the system with haptic proxy. The results are shown in **Table 13**.

The results show that our hypothesis H2 is valid for all the experiments meaning the participants view the system with haptic proxy as more believable and

Section	System Competence		System Usability		Believabitiy	
	Mean	STD	Mean	STD	Mean	STD
Experiment1 with controller	83.3333	14.7939	80.3819	12.0976	41.6667	27.9791
Experiment1 with proxy haptics	75.9259	16.1247	80.5556	15.0581	76.3889	18.0604
Experiment2 with controller	81.9444	13.0020	78.6458	16.1893	54.1667	31.0791
Experiment2 with proxy haptics	86.5741	9.6103	86.8055	7.5552	76.3889	24.0562
Experiment3 with controller	71.2963	12.7202	73.9583	13.6902	50	32.5669
Experiment3 with proxy haptics	84.2593	12.7202	81.2500	17.0644	76.3889	20.6685

Table 12.
Questionnaire scores SED stands for standard error of differences. The responses are from the 12 participants.

Experiments	Result			
	P	t	df	SED
Experiment 1	0.0015	3.6119	22	9.613
Experiment 2	0.0629	1.9587	22	11.345
Experiment 3	0.0270	2.3700	22	11.135

Table 13.
Unpaired t-test results from 12 participants.

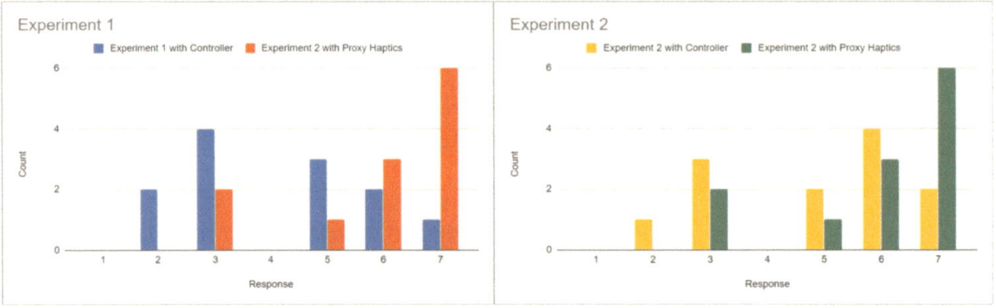

Figure 10.
Responses for the question"I feel no difference between the right- and left-hand manipulations".

closer to the real world than the pure VR version. Also, based on **Figure 10**, users appeared to feel less difference between handling the right and left hand using the haptic proxy.

5. Conclusion

When analyzing the data, we found that in most cases, actions using the haptic proxy simulator are slower than just the VR control system. In fact, in 8 out of 10 different simulations for which we could find a distribution, the haptic proxy was slower. This is normal since by adding a proxy, the movement is restricted by the law of physics. On the other hand, with the standard VR system, users are free of motion restrictions and can move the patient's hand or tools anywhere. We often observed users taking the syringe from inside the patient's body to the target because it was a shorter path, which is impossible in the real world. One advantage of haptic proxy is that collision detection is ensured by the law of physics as opposed to other techniques that require computationally expensive collision detection algorithms.

Based on the questionnaire, all tasks were more credible in the virtual world, and as a result, our H2 hypothesis is validated. Moreover, when asked participants about their opinion, out of the nine who responded to this question, 6 preferred haptic proxy over the controller-only system, saying things like it "Glove with haptic proxy feels more natural", and "With the glove with haptic proxy one feels the weight of things". In addition, when asked about how much they liked the HTC Vive controller as a proxy to the syringe, out of 9 who answered this question, 5 had positive feedback about it. One participant said, "Great, did not even notice the first time".

DOI: http://dx.doi.org/10.5772/intechopen.108330

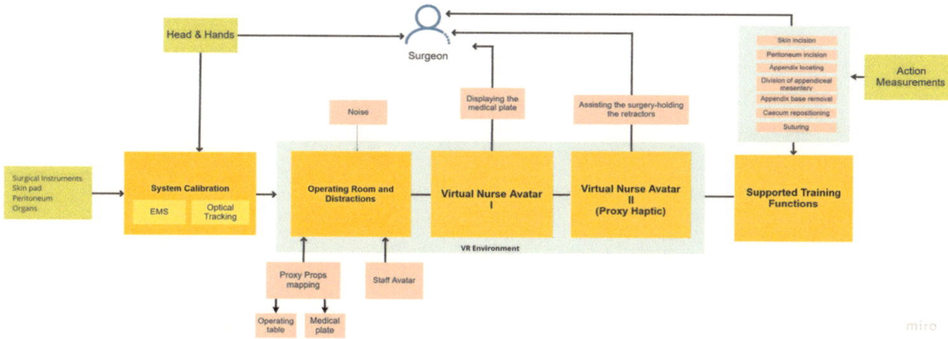

Figure 11.
System diagram of the new appendectomy training system.

In our work, we have attempted to resolve haptic perception in virtual reality toward applications to open surgical training. We did so by real-world mapping objects as proxies for some virtual objects. In our pilot experiment, we used a mannequin as a proxy for a patient and an HTC Vive controller as a proxy for a syringe. The pilot project showed that augmented virtual reality using a haptic proxy could improve the sense of touch for simulations based on virtual reality without needing haptic robots or exotic skeletons. Since the VR-based proxy haptic concept relies heavily on accurately registering virtual and natural objects, occlusion problems were a limiting factor for extending this concept for open surgery simulations. The combination of IK and single-point tracking solved the problem for our simple simulation. Still, they cannot be generalized for complex tasks performed during open surgeries where occlusion will be a significant issue.

From the lessons learned, a new open appendectomy surgical training system is currently under development. In this training system, a combination of fused optical/magnetic trackers and digitizing gloves with pressure sensors will be used to determine the surgeon's body (head and hands) and the 6D position and orientation of the proxy props. In addition, the combination of optical and magnetic tracking will allow us to address the occlusion problem and track the small tools used during the intervention. **Figure 11** shows the proposed block diagram of the new appendectomy simulator. Much work remains to be done to develop a fully functioning open surgery simulator. Still, these pilot studies suggest that using the haptic proxy concept can help achieve this goal.

Author details

Pierre Boulanger*, Thea Wang and Mahdi Rahmani Hanzaki
Department of Computing Science, University of Alberta, Edmonton, Alberta, Canada

*Address all correspondence to: pierreb@ualberta.ca

References

[1] St-John A, Caturegli I, Kubicki N, Kavic S. The rise of minimally invasive surgery: 16 year analysis of the progressive replacement of open surgery with laparoscopy. Society of Laparoscopic and Robotic Surgeons. 2020;**24**(4):1-5

[2] Bhagat KK, Liou WK, Chang CY. A cost-effective interactive 3D virtual reality system applied to military live firing training. Virtual Reality. 2016; **20**(2):127-140

[3] Hamilton MF. Virtual reality: Can it improve the PICU experience? Pediatric critical care medicine—Society of. Critical Care Medicine. 2019;**20**(6): 587-588

[4] Ström P, Kjellin A, Hedman L, Johnson E, Wredmark T, Felländer-Tsai L. Validation and learning in the Procedicus KSA virtual reality surgical simulator. Surgical Endoscopy and Other Interventional Techniques. 2003;**17**(2): 227-231

[5] Seymour NE, Gallagher AG, Roman SA, Obrien MK, Bansal VK, Andersen DK, et al. Virtual reality training improves operating room performance: Results of a randomized, double-blinded study. Annals of Surgery. 2002;**236**(4):458

[6] Van Sickle K, McClusky D III, Gallagher A, Smith C. Construct validation of the ProMIS simulator using a novel laparoscopic suturing task. Surgical Endoscopy and Other Interventional Techniques. 2005;**19**(9): 1227-1231

[7] Duffy A, Hogle N, McCarthy H, Lew J, Egan A, Christos P, et al. Construct validity for the LAPSIM laparoscopic surgical simulator. Surgical Endoscopy and Other Interventional Techniques. 2005;**19**(3):401-405

[8] Roberts KE, Bell RL, Duffy AJ. Evolution of surgical skills training. World Journal of Gastroenterology: WJG. 2006;**12**(20):3219

[9] Huber T, Paschold M, Hansen C, Wunderling T, Lang H, Kneist W. New dimensions in surgical training: Immersive virtual reality laparoscopic simulation exhilarates surgical staff. Surgical Endoscopy. 2017;**31**(11): 4472-4477

[10] Pulijala Y, Ma M, Ayoub A. VR surgery: Interactive virtual reality application for training oral and maxillofacial surgeons using oculus rift and leap motion. In: Serious Games and Edutainment Applications. Cham: Springer; 2017. pp. 187-202

[11] Pulijala Y, Ma M, Pears M, Peebles D, Ayoub A. Effectiveness of immersive virtual reality in surgical training—A randomized control trial. Journal of Oral and Maxillofacial Surgery. 2018;**76**(5): 1065-1072

[12] Lombard M, Ditton T. At the heart of it all: The concept of presence. Journal of Computer-mediated Communication. 1997;**3**(2):JCMC321

[13] Ganni S, Li M, Botden SM, Nayak SR, Ganni BR, Rutkowski AF, et al. Virtual operating room simulation setup (VORSS) for procedural training in minimally invasive surgery–a pilot study. Indian Journal of Surgery. 2020; **2020**:1-7

[14] Perret J, Vander PE. Touching virtual reality: A review of haptic gloves. In: ACTUATOR 2018; 16th International

Conference on New Actuators. VDE. Bremen, Germany: IEEE Xplore; 2018. pp. 1-5

[15] Tiboni M, Borboni A, Verite F, Bregoli C, Amici C. Sensors and actuation technologies in exoskeletons. A Review. Sensor. 2022;**22**(3):1-61

[16] Pacchierotti C, Sinclair S, Solazzi M, Frisoli A, Hayward V, Prattichizzo D. Wearable haptic systems for the fingertip and the hand: Taxonomy, review, and perspectives. IEEE Transactions on Haptics. 2017;**10**(4): 580-600

[17] Srinivasan MA, LaMotte RH. Tactual discrimination of softness. Journal of Neurophysiology. 1995;**73**(1):88-101

[18] Adams MJ, Johnson SA, Lefèvre P, Lévesque V, Hayward V, André T, et al. Finger pad friction and its role in grip and touch. Journal of The Royal Society Interface. 2013;**10**(80):20120467

[19] Keef CV, Kayser LV, Tronboll S, Carpenter CW, Root NB, Finn N III, et al. Virtual texture generated using elastomeric conductive block copolymer in a wireless multimodal haptic glove. Advanced Intelligent Systems. 2020;**2** (4):1-8

[20] Oskouie MA, Boulanger P. Using proxy haptic for a pointing task in the virtual world: A usability study. In: International Conference on Augmented Reality, Virtual Reality and Computer Graphics. Santa Maria al Bagno, Italy: Springer; 2019. pp. 292-299

[21] Simeone AL, Velloso E, Gellersen H. Substitutional reality: Using the physical environment to design virtual reality experiences. In: Proceedings of the 33rd Annual ACM Conference on Human Factors in Computing Systems. Seoul Republic of Korea; 2015. pp. 3307-3316

[22] Henderson SJ, Feiner S. Opportunistic controls: Leveraging natural affordances as tangible user interfaces for augmented reality. In: Proceedings of the 2008 ACM Symposium on Virtual Reality Software and Technology. Bordeaux France; 2008. pp. 211-218

[23] Xu T, An D, Jia Y, Yue Y. A review: Point cloud-based 3D human joints estimation. Sensor. 2021;**21**(5):1-30

[24] Caserman P, Garcia-Agundez A, Konrad R, Steinmetz R. Real-time body tracking in virtual reality using a Vive tracker. Virtual Reality. 2019;**23**(2): 155-168

[25] Tukey JW. Exploratory Data Analysis. Reading, MA: Addison Wesley Publishing; 1977

Chapter 8

Conjugated 3D Virtual Reality Worlds in Spacecraft Attitude Control

Pavel M. Trivailo and Hirohisa Kojima

Abstract

The chapter aims to contribute to the application of virtual reality (VR) in spacecraft attitude control for systems, possessing "Inertial Morphing (IM)" capabilities. The concept of IM, proposed/applied by authors in 2017, is attractive in design of the fully autonomous future space missions, as swift control, requiring minimized energy and computations, can be achieved with exiguous/paltry morphings. To assist rapid planning of the optimized maneuver scenarios, we propose collocated merging of the various VR simulation worlds for the same spacecraft or merging of the VR worlds for the same system (but in different configurations). This enabled concurrent utilization of different methods of modeling, including Poinsot's and Bine's construction and Euler's equations. Therefore, superimposed VR worlds are called "conjugated." We present the classical methods of modeling of the torque-free systems and then show their implementation in the developed VR-integrated interactive package. Effectiveness of the VR-conjugated environment is illustrated with its use for planning of spacecraft de-tumbling, spacecraft 180-degrees inversion, and 90-degrees inversion. The developed VR environment enables utilization of both the body-axes spacecraft coordinate system and inertial coordinate system with instant transition from one into another, switching on/off various virtual reality worlds for multiple supports in the process of mission design.

Keywords: virtual reality, spacecraft, attitude control, area moments of inertia, inertial morphing, inertia ellipsoid, angular momentum sphere, kinetic energy ellipsoid, Poinsot's construction, polhode, herpolhode, separatrix, de-tumbling, spacecraft inversions

1. Introduction

With the onset of computers and development of advanced computer graphics, virtual reality proved to be a very useful and powerful assistant in various areas of human life, including education, professional activities, and entertainment [1]. It may involve virtual and hybrid reality systems, where plain virtual reality (VR) is an environment that is fully digitized and contains no physical elements and hybrid reality (HR) is mostly a virtualized environment; however, some objects have physical existence. By placing trackers on real-world objects, they can be represented and manipulated inside the hybrid environment while providing tactile cues that are not present in virtual reality simulations [2].

Often, VR software, packages, and hardware aim to focus on achieving high graphical and physical realism or aim to add different levels of immersion, in some cases creating a feeling of actual presence in the virtual world [3]. VR and animations are becoming popular in education, enabling explanation of complex principles, simulated scenarios, and interpretation of results. In medicine, VR can provide effective skill transfer into the operating room or remotely [4]. In space flight contexts, "real-life training", which is often expensive, demanding for large facilities or even impossible, VR is used to efficiently extend mission control specialists and astronauts training possibilities [5, 6].

Despite wide embrace of VR application areas, literature on the specialized VR tools in relation to the spacecraft attitude dynamics is quite limited. In the most of these specialized application cases, VR is used as a graphical visualization tool for two main purposes: (1) to assist the mission professionals to visualize orbits and/or spacecraft systems externally (illustrated in [7]) or (2) to assist in space crew mission training, helping to understand the local spacecraft or station environment [8]. These two main areas, essentially, are enhancing understanding of the inertial or non-inertial simulated environments.

The current work aims to contribute to the application of VR in spacecraft attitude dynamics, in particular, to the systems, enabling "Inertial Morphing (IM)": IM with exiguous/paltry number of IM control actions is a new concept, proposed by the authors in 2017 and applied to the spacecraft attitude control [9, 10]. The exiguous control enables swift control, requiring minimized energy and computations, and allows design of the fully autonomous future space missions [11, 12].

This work involves collocated merging of the various virtual reality simulation worlds for the same spacecraft system (where VR worlds are used to illustrate different methods of modeling) or involves merging of the virtual reality worlds for the same system but in different configurations (in this case, the corresponding VR worlds are to be simultaneously used in the development of the integrated spacecraft attitude maneuver scenario). Therefore, these superimposed VR worlds are called "conjugated". We present the theory behind the developed VR-integrated interactive environment and then the applications of the programmed VR environment, based on the presented theory. Use of the VR-conjugated environment is illustrated with its use for planning of a few illustration attitude maneuvers, including spacecraft de-tumbling, spacecraft 180-degrees inversion, and 90-degrees inversion [13]. The developed VR environment enables utilization of both, the body-axes spacecraft coordinate system, and inertial coordinate system with instant transition from one into another, switching on/off various virtual reality worlds for multiple supports in the design process [10].

2. Main conventions and abbreviations in this chapter

It would be appropriate to declare some of the main conventions to be used in this work. Firstly, SI units will be employed; therefore, all area moments of inertia will be in $kg \times m^2$, angular velocity and its components will be in rad/s, kinetic energy will be in Joules ($kg \times m^2 \times s^{-2}$), angular momentum will be in $kg \times m^2 \times s^{-1}$ (which is the same as N·m·s), and these units for brevity will be omitted everywhere in the text after the numerical values.

Also, we will often use abbreviation KEE, AME, and AMS for the "kinetic energy ellipsoid", "angular momentum ellipsoid", and "angular momentum sphere", VR for the "virtual reality". In terms of notations, for brevity, we will use $[I]$ and $[\omega]$ to express components of the principal area moments of inertia and angular velocity of the system.

3. "Poinsot's Construction": graphical method in attitude kinematics

3.1 Poinsot's Construction

Various spacecraft systems can be modeled as torque-free systems. For simulation of their motions, including regular spin, flipping, or tumbling, Euler's equations can be used. This method, involving analytics and numerical methods, will be employed in the later sections. However, basic explanation of the motions can be explained even without solving the Euler's equations, but just employing the geometrical interpretation of the torque-free motion. This elegant method of representing kinematically the motion of a body is known in classical mechanics, as "Poinsot's construction" (after Louis Poinsot, who published the method in 1834 [14]).

Paying tribute to his amazing discoveries, we wish to present the basic information on this wonderful scientist. Louis Poinsot (January 3, 1777–December 5, 1859) was a French mathematician and physicist. Being passionate about abstract mathematics, he invented the geometrical mechanics. For his numerous contributions, he was elected Fellow of the Royal Society of London in 1858. He died in 1859 at the age of 82. His is one of the 72 names inscribed on the Eiffel Tower: where names of the French scientists, engineers, and mathematicians are engraved in recognition of their contributions [15]. A related fragment from the Eiffel Tower is shown in **Figure 1**. Symbolically that Poinsot's name is next to Focault's name: Poinsot was a contemporary of Léon Foucault, who invented the gyroscope and whose pendulum experiments provided incontrovertible evidence that the Earth rotates.

Poinsot's construction will be shortly reviewed in this section. Torque-free motion is subject to four constants: the kinetic energy of the body and the three components of the angular momentum, expressed in the *inertial* coordinate system. The angular velocity vector $\boldsymbol{\omega}$ of the rigid rotor is not constant, but satisfies Euler's equations. Without explicitly solving these equations, Louis Poinsot was able to visualize the motion of the endpoint of the angular velocity vector. He used the conservation of

Figure 1.
Name of Louis Poinsot on Eiffel Tower: (a) the location of the names on the Tower [16]; (b) a fragment of plaque on the South-East side of the Tower with Poinsot's name [17].

kinetic energy and angular momentum as two constraints on the motion of the angular velocity vector $\boldsymbol{\omega}$.

3.2 Poinsot's Kinetic Energy Ellipsoid

Let us first explore the kinetic energy K constraint, writing its analytical expression:

$$K = \frac{1}{2} I_{xx}\omega_x^2 + \frac{1}{2} I_{yy}\omega_y^2 + \frac{1}{2} I_{zz}\omega_z^2 = const = K_0 \tag{1}$$

This can be written as

$$I_{xx}\omega_x^2 + I_{yy}\omega_y^2 + I_{zz}\omega_z^2 = 2K_0 \tag{2}$$

which represents an ellipsoid, often called "Poinsot's ellipsoid (PE)". It should not be mixed up with the "standard inertial ellipsoid", which differs from the Poinsot's ellipsoid by the scale factor $\sqrt{2K}$ [18]. However, we further re-write (2) in more convenient, non-dimensional form:

$$(\overline{\omega}_x/a)^2 + (\overline{\omega}_y/b)^2 + (\overline{\omega}_z/c)^2 = 1 \tag{3}$$

where

$$\overline{\omega}_x = \omega_x/\sqrt{2K_0}, \quad \overline{\omega}_y = \omega_y/\sqrt{2K_0}, \quad \overline{\omega}_z = \omega_z/\sqrt{2K_0} \text{ and} \tag{4}$$

$$a = 1/\sqrt{I_{xx}}, \quad b = 1/\sqrt{I_{yy}}, \quad c = 1/\sqrt{I_{zz}} \tag{5}$$

Eq. (3) is a standard form representation of a 3D ellipsoid, called "Poinsot's kinetic energy ellipsoid (KEE)", which is shown with its notations in **Figure 2a**. Each quantity a, b, c is equal to the half of the length of the corresponding principal axis of the ellipsoid. If two of the axes have the same length, then the ellipsoid is an ellipsoid of revolution, also called a spheroid. In this case, the ellipsoid is invariant under a rotation around the third axis, and there are thus infinitely many ways of choosing the two perpendicular axes of the same length. If the third axis is shorter, the ellipsoid is

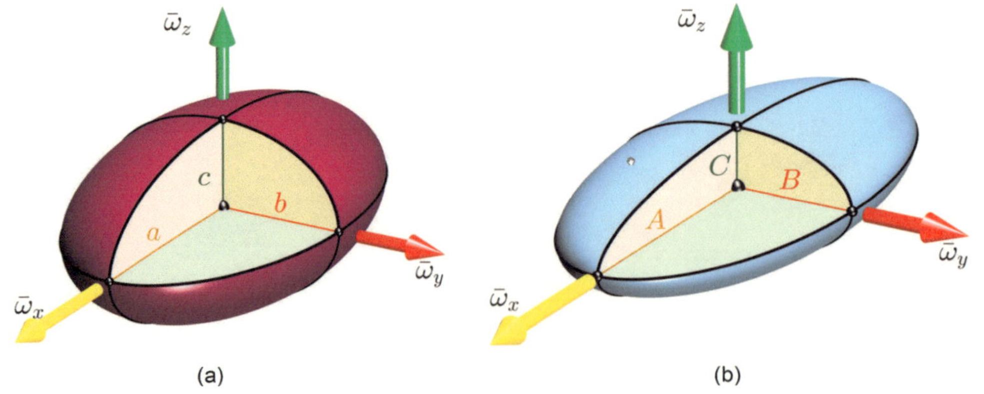

Figure 2.
(a) Poinsot's kinetic energy ellipsoid; (b) Poinsot's angular momentum ellipsoid.

an oblate spheroid; if it is longer, it is a prolate spheroid. If the three axes have the same length, the ellipsoid is a sphere.

3.3 Poinsot's Angular Momentum Ellipsoid

Similar to K, the angular momentum H is another constraint of the system, as for the torque-free motion case has constant squared length, which can be expressed in terms of angular velocity components as follows:

$$|\mathbf{H}|^2 = (I_{xx}\omega_x)^2 + \left(I_{yy}\,\omega_y\right)^2 + (I_{zz}\,\omega_z)^2 = const = H_0 \tag{6}$$

This relationship can be also re-written in terms of the same non-dimensional quantities $\overline{\omega}_x,\ \overline{\omega}_y,$ and $\overline{\omega}_z$ as follows:

$$(\overline{\omega}_x/A)^2 + \left(\overline{\omega}_y/B\right)^2 + (\overline{\omega}_z/C)^2 = 1 \tag{7}$$

where

$$A = H_0/\left(I_{xx}\sqrt{2K}\right),\quad B = H_0/\left(I_{yy}\sqrt{2K}\right),\quad C = H_0/\left(I_{zz}\sqrt{2K}\right) \tag{8}$$

Eq. (7) represents another 3D ellipsoid, called "angular momentum ellipsoid" (AME), or "momental ellipsoid", or "ellipsoid of inertia". With its notations, it is shown in **Figure 2b**. Note, that the polar reciprocal of AME with regard to its center is another ellipsoid, which is sometimes called the ellipsoid of gyration [19].

3.4 Poinsot's angular velocity polhodes

After the main notations were introduced, let us consider a particular illustration case of the system with the following parameters: $\left[I_{xx},\ I_{yy},\ I_{zz}\right] = [2,\ 3,\ 6]$ and $\left[\omega_x,\ \omega_y,\ \omega_z\right] = [0.1,\ 8,\ 1]$. Then, system's constants are $K = 99.01$ and $H_0 = 24.74$ and the key characteristics of the KEE and AME are:
$[a,\ b,\ c] = [0.7071,\ 0.5774,\ 0.4082];[A,\ B,\ C] = [0.8790, 0.5860, 0.2930]$.

We use the VR tools and show the corresponding KEE and AME surfaces separately in **Figure 3a** and **3b**. In the figure, we also display $\overline{\boldsymbol{\omega}}$ (scaled vector of the angular velocity) and $\overline{\mathbf{H}}$ (non-dimensional vector of the angular momentum). The scaling factor of 2 is applied to $\overline{\boldsymbol{\omega}}$ to ensure that the vector is visible, otherwise it would be completely hidden inside of the KEE surface. So, instead of $\overline{\boldsymbol{\omega}}$ to enable visualization we show $2 \times \overline{\boldsymbol{\omega}}$.

In **Figure 3c** we show KEE and AME ellipsoids as merged and co-centered. **Figure 3c** then instantly reveals that after completing collocation, two ellipsoids intersect along two lines. For better observation of the intersecting surfaces, we also show in **Figure 3d** the AME as semi-transparent object. This last figure graphically illustrates data in Eq. (3) and Eq. (6), in particular, the difference in the values of a and A, that is, values of the semi-major axes of the KEE and AME.

It can be observed from **Figure 3c** that the tip of the vector $\overline{\boldsymbol{\omega}}$ resides on one of the polhodes. In fact, the angular velocity vector $\boldsymbol{\omega}$ is tracing out the path, called the polhode, which is generally circular or taco-shaped. In the fashion of the day, Poinsot coined the terms polhode and its counterpart, herpolhode, to describe this wobble in the motion of rotating rigid bodies. Poinsot derived these terms from the ancient

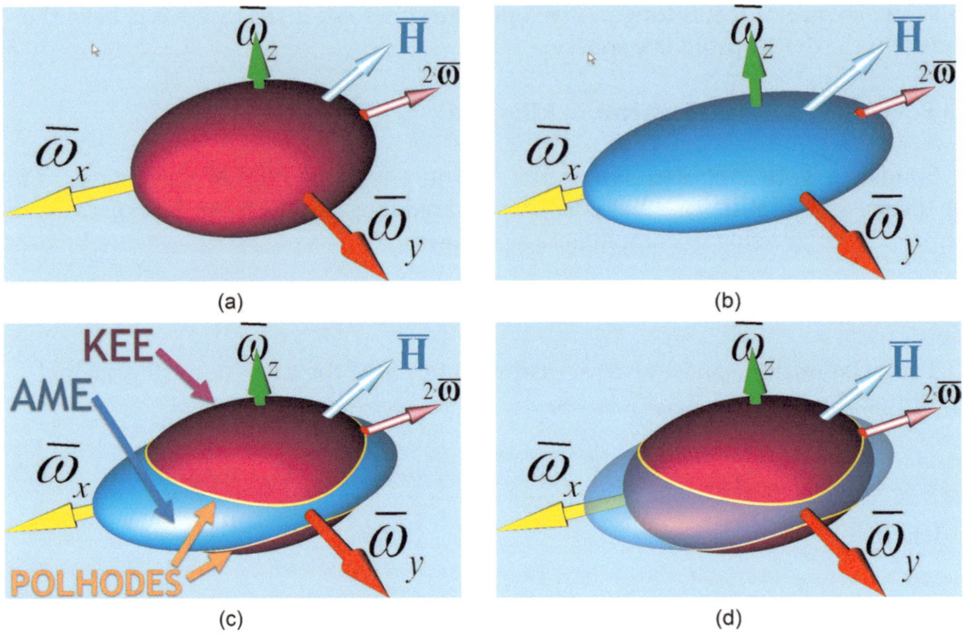

Figure 3.
Illustration of polhodes: (a) KEE; (b) AME; (c) merging KEE and AME via *co-centering reveals polhodes; (d) for better visualization, AME is shown semi-transparent.*

Greek πόλος (pólos—pivot *or* end of an axis) + ὁδός (hodós—path *or* way)—thus, polhode is the "path of the pole" [20].

If the rotating rigid body is symmetric (has two equal moments of inertia), the vector **ω** slides along the sides of a cone (and its endpoint draws a circle). This is known as the torque-free precession of the rotation axis of the rotor.

An observer, attached to the body frame coordinate system, sees the angular velocity vector $\overline{\boldsymbol{\omega}}$ as sliding along the surface of a cone, called the body cone, whose intersection with the inertia ellipsoid is the polhode. The observer also sees that the vector of the angular velocity is generally circling around one of the principal axes, as illustrated in **Figure 4** with three representative cases. In a particular special case of a symmetrical body, the inertia ellipsoid is an ellipsoid of revolution, so that the polhode becomes a circle perpendicular to the spin axis for the coning vector $\overline{\boldsymbol{\omega}}$.

In this work, a special attention will be given to another class of other special cases, when $H^2 \approx 2K_0 I_{int}$, where I_{int} denotes the intermediate value of the moment of inertia. In these cases, the Poinsot's polhodes shapes start getting "taco" shapes (shown in

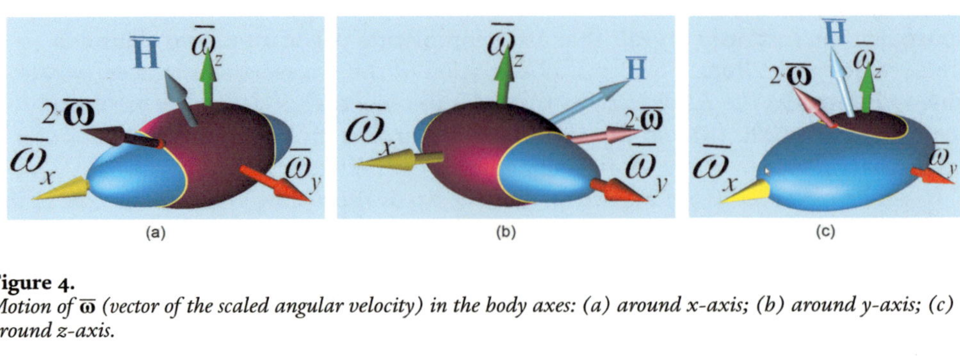

Figure 4.
Motion of $\overline{\boldsymbol{\omega}}$ *(vector of the scaled angular velocity) in the body axes: (a) around x-axis; (b) around y-axis; (c) around z-axis.*

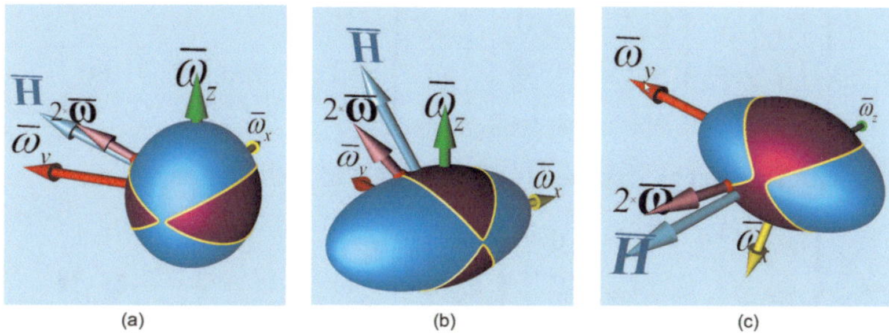

Figure 5.
"Taco"-shapes of the Poinsot's polhodes for the cases in which $H^2 \approx 2 K_o I_{int}$.

Figure 5) while approaching the separatrices and the system motion would represent unstable flipping along the intermediate body axis, if the system is initially provided with significant spin about the intermediate axis.

Polhode can be described analytically and without loss of generality, we perform derivation of the associated equation, assuming that $a < b < c$. We consider the following parametrization for the xy plane and substitute it into Eq. (7):

$$\begin{cases} \overline{\omega}_x = A\rho\cos\theta \\ \overline{\omega}_y = B\rho\sin\theta \end{cases} \Rightarrow \quad \overline{\omega}_z^2 = C^2(1-\rho^2) \tag{9}$$

Substitution into Eq. (3) gives:

$$\left(\frac{A\rho\,\cos\theta}{a}\right)^2 + \left(\frac{B\rho\,\cos\theta}{b}\right)^2 + \frac{C^2\,(1-\rho^2)}{c^2} = 1 \tag{10}$$

which results in the parametric solution for ρ:

$$\rho = \pm\sqrt{\left(1-\frac{C^2}{c^2}\right)/\left(\frac{A^2}{a^2}\cos^2\theta + \frac{B^2}{b^2}\sin^2\theta - \frac{C^2}{c^2}\right)} \tag{11}$$

With this equation, for each value of θ, the corresponding value of ρ can be calculated and then the values of x, y, and z can be finally determined, using Eqs. (9). Alternatively, all Eqs. (9) can be expressed in terms of the single angular parameter θ:

$$\begin{aligned} \overline{\omega}_x &= \pm\frac{A\cos\theta}{a}\sqrt{\left(1-\frac{C^2}{c^2}\right)/\left(\frac{A^2}{a^2}\cos^2\theta + \frac{B^2}{b^2}\sin^2\theta - \frac{C^2}{c^2}\right)}; \\ \overline{\omega}_y &= \pm\frac{B\sin\theta}{b}\sqrt{\left(1-\frac{C^2}{c^2}\right)/\left(\frac{A^2}{a^2}\cos^2\theta + \frac{B^2}{b^2}\sin^2\theta - \frac{C^2}{c^2}\right)}; \\ \overline{\omega}_z &= \pm\frac{C}{c}\sqrt{\left(1-\frac{C^2}{c^2}\right)/\left(\frac{A^2}{a^2}\cos^2\theta + \frac{B^2}{b^2}\sin^2\theta - \frac{C^2}{c^2}\right)}. \end{aligned} \tag{12}$$

3.5 Poinsot's Herpolhode and Invariable Plane

The angular momentum vector **H** can be expressed in terms of the moment of inertia tensor **I** and the angular velocity scaled vector $\overline{\boldsymbol{\omega}}$:

$$\mathbf{H} = \begin{Bmatrix} H_x \\ H_y \\ H_z \end{Bmatrix} = \begin{bmatrix} I_{xx} & 0 & 0 \\ 0 & I_{yy} & 0 \\ 0 & 0 & I_{zz} \end{bmatrix} \begin{Bmatrix} \omega_x \\ \omega_y \\ \omega_z \end{Bmatrix} = \mathbf{I} \cdot \boldsymbol{\omega} = \sqrt{2K_0}(\mathbf{I} \cdot \overline{\boldsymbol{\varpi}}) \tag{13}$$

Similarly, kinetic energy can be also expressed as follows:

$$K_0 = \frac{1}{2} \begin{Bmatrix} \omega_x \\ \omega_y \\ \omega_z \end{Bmatrix} \begin{bmatrix} I_{xx} & 0 & 0 \\ 0 & I_{yy} & 0 \\ 0 & 0 & I_{zz} \end{bmatrix} \begin{Bmatrix} \omega_x \\ \omega_y \\ \omega_z \end{Bmatrix} = \frac{1}{2} \boldsymbol{\omega} \cdot \mathbf{I} \cdot \boldsymbol{\omega} = \frac{\sqrt{2K_0}}{2} \overline{\boldsymbol{\varpi}} \cdot \mathbf{H} \tag{14}$$

Eq. (14) then leads to the following:

$$\overline{\boldsymbol{\varpi}} \cdot \mathbf{H} = \sqrt{2K_0} \tag{15}$$

The dot product of two vectors can be calculated, using (a) projection of $\overline{\boldsymbol{\varpi}}$ on $\mathbf{H}$, giving result $\overline{\boldsymbol{\varpi}}_{\mathbf{H}} \times |\mathbf{H}| = H_0\, \overline{\boldsymbol{\varpi}}_{\mathbf{H}}$ or (b) projection of $\mathbf{H}$ on $\overline{\boldsymbol{\varpi}}$, giving result $|\overline{\boldsymbol{\varpi}}| \times \mathbf{H}_{\overline{\boldsymbol{\varpi}}}$. Both methods lead to the constant value of $\sqrt{2K_0}$; however, method (a) is more preferred, as involves a product of two constants, value of the angular momentum (which is constant due to the "Law of Conservation of angular momentum" for the torque-free system), and, hence, the constant projection $d = \overline{\boldsymbol{\varpi}}_{\mathbf{H}}$ of the angular velocity scaled vector $\overline{\boldsymbol{\varpi}}$ on $\mathbf{H}$, shown as OD in **Figure 6**. This choice enabled Poinsot to establish a remarkable plane p, the *invariable plane*, being perpendicular to the angular momentum $\mathbf{H}$ and at the distance OD from the origin O:

$$d = |OD| = \overline{\boldsymbol{\varpi}}_{\mathbf{H}} = \sqrt{2K_0}/H_0 \tag{16}$$

As the rigid body moves, its inertia ellipsoid (which is locked in the body axes system) revolves around the fixed point O in a peculiar way: It rolls on the invariable plane without slipping, with every new point of contact "painting" two continuous curves: *polhode* on the inertia ellipsoid and *herpolhode* on the invariable plane, as if it was a double-sided carbon paper between the invariable plane and the ellipsoid.

The herpolhode is always concave to the origin, described by Goldstein [18] as "snakelike". The herpolhode is generally an open curve, which means that the rotation does not perfectly repeat, but the polhode is a closed curve.

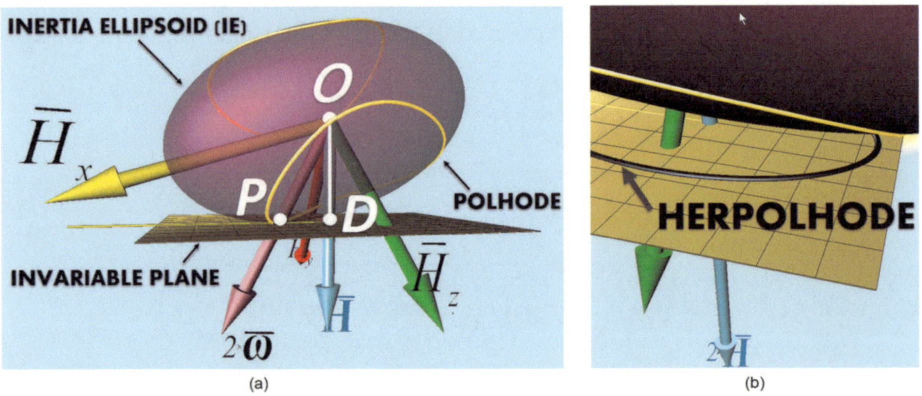

Figure 6.
Invariable plane, normal to the inertia ellipsoid at point P (being a tip of the vector $\overline{\varpi}$) and at the invariable (constant) distance "d" from O: $d = |OD| = \sqrt{2K}/H_o$.

In summary of the presentation of the Poinsot's construction, we see that in an absolute reference frame, the instantaneous angular velocity vector is the point of intersection between a fixed invariable plane and a kinetic-energy ellipsoid that is tangent to it and rolls around on it without slipping.

Vector of the angular momentum for the torque-free systems is fixed in space; however, angular momentum vector is not. Poinsot's construction is useful in understanding on how the vector $\overline{\boldsymbol{\omega}}$ moves. In order to understand the motion of the vector of the angular momentum $\overline{\mathbf{H}}$, another graphical method is to be used. It is based on the Binet's ellipsoid, named after the French mathematician, physicist and astronomer Jacques Philippe Marie Binet (1786–1856).

4. Analysis of the attitude kinematics of the Inertially Morphed Systems

4.1 Concept of inertial morphing

In the previous publications by the authors [11, 21, 22], we proposed for wide range variations of the attitude motions of the spacecraft to deliberately change its principle moments of inertia during the attitude maneuvering flight. We called this concept as "Inertial Morphing (IM)" and envisaged that it can be achieved in many different ways *via* variety of methods/principles, including reposition of control masses with smart mechanisms or *via* components docking/undocking; redistribution of the magnetic control liquids with magnetic fields; rejection of controlled liquids *via* ejection/ablation, etc. Efforts of the IM systems can be also enhanced *via* specifically designed motions of the spacecraft appendages to assist in attitude control. We believe that the novel concept of IM enables design and construction of the inertially morphed spacecraft, possessing acrobatic capabilities, and may allow design of new class of gyroscopic systems with a "sense" of time.

The concept of IM was prompted by the observation of rigid bodies in space performing the intriguing flipping motion. The phenomenon was first observed in space by famous US scientist-astronaut Owen Kay Garriott on-board Skylab in 1973, but later was publicized as "Dzhanibekov's effect" named after V. Dzhanibekov, who observed the flipping of a spinning wing nut during his space flight in 1985 [23]. Interestingly, the resulting flipping motions of rigid bodies with various shapes can be explained by the L.Euler's equations, derived in 1785 for rigid bodies with *fixed* area moments of inertia.

We wish to utilize and expand the functionality of the Poinsot's graphical method. As per the IM concept, we will be specifically emphasizing on the variation of the principal moments of inertia (i.e., implying that $[\mathbf{I}] = [\mathbf{I}(t)]$), not initially envisaged and explored by L. Poinsot and L. Euler in their works, we will need to treat moments of inertia as variables, expand classical Euler's equation of motion, and involve another graphical construction, based on Binet's construction, which is mostly suitable for the analysis of morphed systems, as is centered around the angular momentum, involving $[\mathbf{I}]$.

4.2 Definition of the AMS and Binet's KEE

For the future analysis and interpretation of results, let us introduce the following non-dimensional coordinates $\overline{H}_x$, $\overline{H}_y$ and $\overline{H}_z$, associated with x, y, and z body axes:

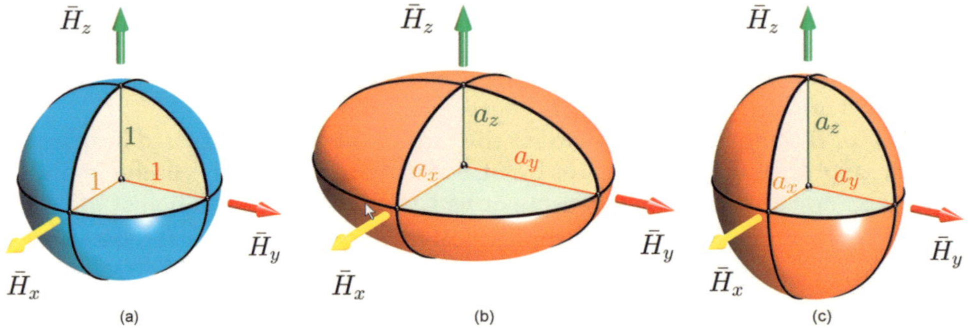

Figure 7.
Surfaces for attitude dynamics: (a) unit radius angular momentum sphere (AMS) in the non-dimensional coordinates; (b) example of kinetic energy ellipsoid (KEEs) for the system with $a_x = 1.2$; $a_y = 1.4$; $a_z = 0.8$; *(c) KEE for* $a_x = 0.8$; $a_y = 1$; $a_z = 1.2$ *case.*

$$\begin{aligned}&\overline{H}_x(t) = H_x/H_0; \quad \overline{H}_y(t) = H_y/H_0; \quad \overline{H}_z(t) = H_z/H_0,\\ &\text{where } H_x = I_{xx}\omega_x; \quad H_x = I_{xx}\omega_x; \quad H_x = I_{xx}\omega_x\\ &\text{and } H_0 = \sqrt{H_x^2 + H_y^2 + H_z^2} = \sqrt{(I_{xx}\omega_x)^2 + (I_{yy}\omega_y)^2 + (I_{zz}\omega_z)^2}\end{aligned} \tag{17}$$

We assume that the spacecraft system is in free flight. Then the fundamental "Law of Conservation of angular momentum" can be applied to the system and this law can expressed in a compact form:

$$\overline{H}_x^{\,2} + \overline{H}_y^{\,2} + \overline{H}_z^{\,2} = 1 \tag{18}$$

This equation can be interpreted graphically as a unit sphere (i.e., simply a sphere of radius one), shown in **Figure 7a**.

Kinetic energy of the system can be written in terms of the area moments of inertia and components of the angular velocity:

$$K(t) = \frac{1}{2}(I_{xx}\omega_x)^2 + \frac{1}{2}(I_{yy}\omega_y)^2 + \frac{1}{2}(I_{zz}\omega_z)^2 \tag{19}$$

This expression can be further re-written as follows:

$$K(t) = \left[\frac{H_x(t)}{\sqrt{2\ I_{xx}(t)}}\right]^2 + \left[\frac{H_y(t)}{\sqrt{2\ I_{yy}(t)}}\right]^2 + \left[\frac{H_z(t)}{\sqrt{2\ I_{zz}(t)}}\right]^2.$$

In view of the introduced non-dimensional coordinates $\overline{H}_x$, $\overline{H}_y$, and $\overline{H}_z$, expression for $K(t)$, similar to H_0, can be also expressed in the concise form:

$$\left(\frac{\overline{H}_x}{a_x}\right)^2 + \left(\frac{\overline{H}_y}{a_y}\right)^2 + \left(\frac{\overline{H}_z}{a_z}\right)^2 = 1 \tag{20}$$

where

$$a_x = \frac{\sqrt{2K(t)I_{xx}(t)}}{H_0}; \quad a_y = \frac{\sqrt{2K(t)I_{yy}(t)}}{H_0}; \quad a_z = \frac{\sqrt{2K(t)I_{zz}(t)}}{H_0} \tag{21}$$

Eq. (20) corresponds to the so-called Binet's KEE [18], also fixed to the body axes non-inertial coordinate system, as Poinsot's AME and KEE ellipsoids.

However, Binet's KEE, given by Eq. (20) is constructed in the non-dimensional coordinates $\overline{H}_x$, $\overline{H}_y$, and $\overline{H}_z$, and has the semi-major axes, equal to a_x, a_y, and a_z. In the applications of the IM concept, any combinations of a_x, a_y, and a_z may occur. In **Figure 7b** and **7c**, we present two contrast cases of the kinetic energy ellipsoids (KEEs). In the first example, $a_z < a_x < a_y$, however in the second example, the ascending order of the values of the semi-major axes is different: $a_x < a_y < a_z$. The axis, associated with the smallest (out of a_x, a_y and a_z) value, will be called "minimum axis" of inertia. The axis, associated with the largest (out of a_x, a_y and a_z) value, will be called "maximum axis" of inertia and the third axis will be called "intermediate axis" of inertia. Special attention should be paid to the identification of the intermediate axis, as if the system is provided with the spin about this axis, the resultant motion will be unstable and the system would start flipping motion. In relation to two cases in **Figure 7b** and **7c**, for initiation of the unstable flips, the system (*b*) should be provided with the predominant spin about the x axis, whereas the second system would start flips if the predominant spin is provided about the y axis.

4.3 Co-centering AMS and KEE in a numerical example

Let us consider an illustrative numerical example of the system with the following moments of inertia and initial components of the angular velocity:

$$I_{xx} = 2; \quad I_{yy} = 4; \quad I_{zz} = 6; \quad \omega_{x0} = 0.1; \quad \omega_{y0} = 1; \quad \omega_{z0} = 1 \qquad (22)$$

These parameters enable to use Eq. (17), and Eq. (19) to determine $H_0 = 7.21\ \text{kg} \times \text{m}^2/\text{s}$ and $K = 5.01\ \text{J}$, then use Eq. (21) to calculate a_x = 0.62; a_y = 0.88; a_z = 1.07 and therefore identify "y" axis as an axis, corresponding to the intermediate area moment of inertia. The corresponding KEE can be constructed and plotted together with the co-centered, collocated AMS (shown as semi-transparent surface), as shown in **Figure 8**. From the figure, we can clearly see that the largest portion of the KEE is

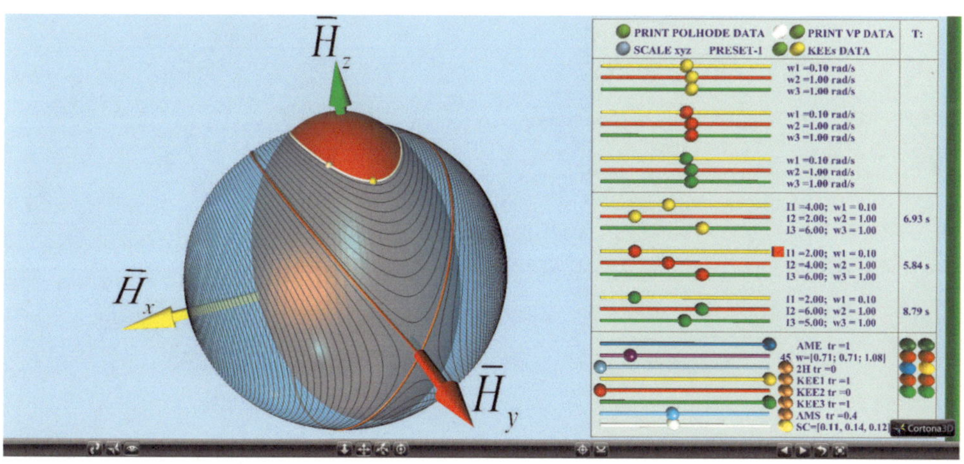

Figure 8.
Co-centered, collocated angular momentum sphere and kinetic energy ellipsoid presented in the virtual reality interface for the illustration case.

inside the AMS; however, top and bottom portions of the KEE are extruding outside the AMS. **Figure 8** enables us to observe two symmetrical intersection lines between KEE and AMS, painted with white color. Note that only the top intersection line, polhode, is seen in the static **Figure 8**, whereas the bottom polhode is obscured with the KEE. Their size and shape reflect the nature of the attitude motion of the system; therefore, polhodes are very important visual integrated identifiers of the rotation of the system. By just viewing the shape, the designer, for example, could instantly determine if the system is in complex tumbling, unstable flips or regular spin, could also determine the "distribution" of rotation between the body axes.

Figure 8 also presents with a red bottom dot an initial state of the analyzed system. Indeed, Eq. (17) can be used to calculate non-dimensional initial coordinates of the system:

$$\overline{H}_{x0} = 0.0277; \quad \overline{H}_{y0} = 0.5545; \quad \overline{H}_{z0} = 0.8317. \tag{23}$$

It would be important to note that there may be infinite number of combinations of other initial conditions, resulting in their initial state points to reside on the same polhode. As an example, we show with the left red dot in **Figure 8** another set of system's initial conditions, placing the state point on the same polhode. Similar to (23), we present the other set of initial conditions with the following numbers:

$$\overline{H}_{x0} = 0.2217; \quad \overline{H}_{y0} = 0.3374; \quad \overline{H}_{z0} = 0.9149. \tag{24}$$

For illustration purpose, in **Figure 9** we represent two different systems in a single combined plot with their co-centered KEEs and identical AMSs. The feature of this case is that they both have identical separatrices, shown with red- and

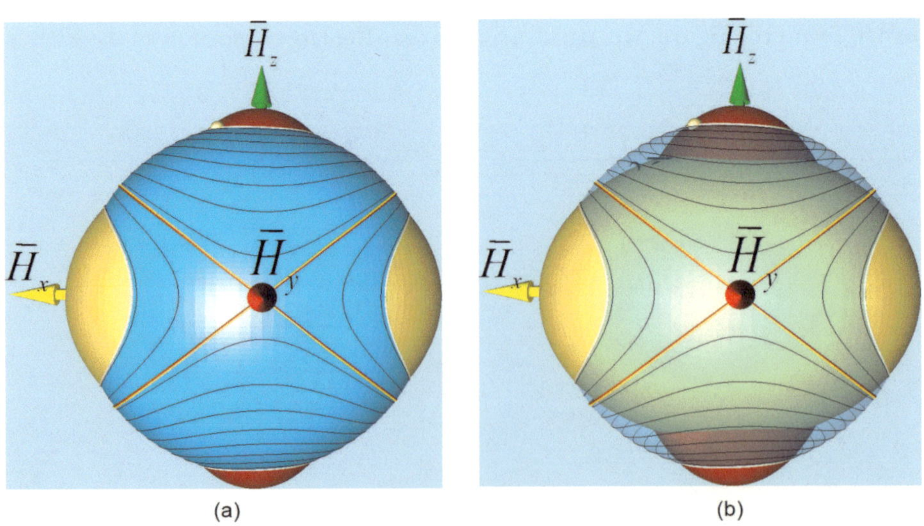

Figure 9.
Two co-centered conjugated systems with $I_A = [3,4,9]$, $\omega_A = [2, \; 1, \; 2]$ *and* $I_B = [8, 6.53, 5]$, $\omega_B = [-15, \; -15, \; -12]$, *which have the same separatrices: (a) the identical AMS shapes are shown as opaque (non-transparent surfaces); (b) the AMSs are shown as semi-transparent, revealing red KEE for the first system and yellow KEE for the second system.*

yellow-coinciding lines. The condition for the separatrices is: $H^2 = 2KI_{yy}$; however, selection of the initial conditions on two sides of the separatrix does not necessary mean that the switch from $H^2 > 2KI_{yy}$ to $H^2 < 2KI_{yy}$ (or *vice versa*) would occur. In this respect, for both, "*A*" and "*B*" systems, the same relationship $H^2 > 2K\, I_{yy}$ is satisfied.

4.4 Torque-free Euler's equations of motion of the rigid body

The famous Euler's equations, describing attitude dynamics of the rigid body, can be written on the matrix form:

$$\begin{bmatrix} I_{xx} & 0 & 0 \\ 0 & I_{yy} & 0 \\ 0 & 0 & I_{zz} \end{bmatrix} \begin{Bmatrix} \dot{\omega}_x \\ \dot{\omega}_y \\ \dot{\omega}_z \end{Bmatrix} = \begin{Bmatrix} \left(I_{yy} - I_{zz}\right) \omega_y \, \omega_z \\ \left(I_{zz} - I_{xx}\right) \omega_z \, \omega_x \\ \left(I_{xx} - I_{yy}\right) \omega_x \, \omega_y \end{Bmatrix} \tag{25}$$

These equations can be solved numerically for any combination of I_{xx}, I_{yy}, I_{zz}, and initial conditions, including, for example, Eq. (23) and/or Eq. (24). Then, the time histories for ω_x, ω_y, and ω_z can be determined, results presented in the non-dimensional coordinates $\overline{H}_x$, $\overline{H}_y$, and $\overline{H}_z$ and eventually plotted as trajectories of the tip of the angular momentum vector $\overline{\mathbf{H}}$. It is remarkable that the solution trajectory would exactly correspond to the polhode, that is, intersection line between the KEE and AMS surfaces. Therefore, one of the aims of the dynamics analyst may be determination of the shape of the polhode of the system, which would enable fast classification of the attitude motion by visual inspection.

By the way, Eq. (25) can be used to determine direction of motion of the vector $\overline{\mathbf{H}}$. For example, considering a red point in **Figure 8**, symbolizing the tip of the angular momentum vector, located in the vicinity of $x = 0$ plane, we set $\omega_x = 0$ and reduce Euler's equations to one equation only:

$$I_{xx}\, \dot{\omega}_x = \left(I_{yy} - I_{zz}\right) \omega_y \, \omega_z \tag{26}$$

Then, sliding slightly the point from the $x = 0$ plane along polhode, to the position, where $\omega_y > 0$ and $\omega_z > 0$ (i.e., to the xyz-positive octant), we determine that $\dot{\omega}_x < 0$, as $I_{yy} < I_{zz}$ for the selected example. Therefore, red dot point must move in *anti-clock direction* around z axis (determined with the right-hand rule), to ensure reduction in the ω_x value with time.

4.5 Analytical expressions for Polhodes

Without the loss of generality, let us assume that $a < b < c$. For this case, the xz cross section of the co-centric AMS and KEE is presented in **Figure 10**.

For the analysis of the intersection of the AMS and KEE along polhodes, we introduce the following parametrization in the xy plane, being an orthogonal plane to the plane of the xz cross section:

$$\overline{H}_x = r \cos\theta; \quad \overline{H}_y = r \sin\theta \tag{27}$$

In Eq. (27), the angle θ and radius r are the polar coordinates in the xy plane. In view of Eq. (27) and Eq. (18), describing the AMS, we can deduct that

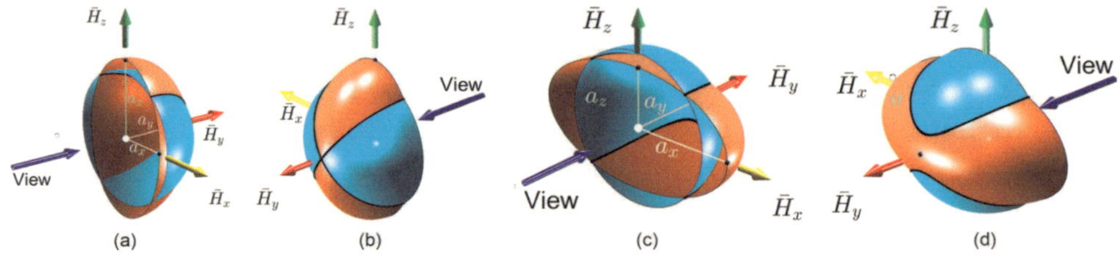

Figure 10.
Cross section of the KEE and AMS with the xz plane for two particular cases of the systems with: (a)-(b) $a_x = 0.8$; $a_y = 1$; $a_z = 1.2$; *(c)-(d)* $a_x = 1.4$; $a_y = 1.02$; $a_z = 0.8$.

$$\overline{H}_z = \sqrt{1 - r^2} \tag{28}$$

Substitution of Eqs. (27) and (28) into Eq. (20) for the KEE enables us to re-write it in terms of the polar coordinates as follows:

$$(r\ \cos\theta/a_x)^2 + (r\ \sin\theta/a_y)^2 + \left(\sqrt{1 - r^2}/a_z\right)^2 = 1 \tag{29}$$

It can be solved for the radius r becoming a function of only angle θ, assuming that the system's dynamics characteristics are all known:

$$r = \pm\sqrt{\frac{1 - (1/a_z)^2}{(\cos\theta/a_x)^2 + (\sin\theta/a_y)^2 - (1/a_z)^2}} \tag{30}$$

Running a cycle for all polar angle θ in the range between 0 and 360 degrees, corresponding values of r can be determined, using Eq. (30). And then, for each matching pair of angle θ and radius r, the triplets of corresponding coordinates $\overline{H}_x$, $\overline{H}_y$, and $\overline{H}_z$ of the polhodes can be determined, using Eqs. (27-28).

Eqs. (27, 28, 30) can be used to plot polhodes in the upper and lower quarters of the AMS for the illustrative system in **Figure 11a**. To plot polhodes in the left and right quarters, in Eq. (30) x- and z-related parameters should be swapped.

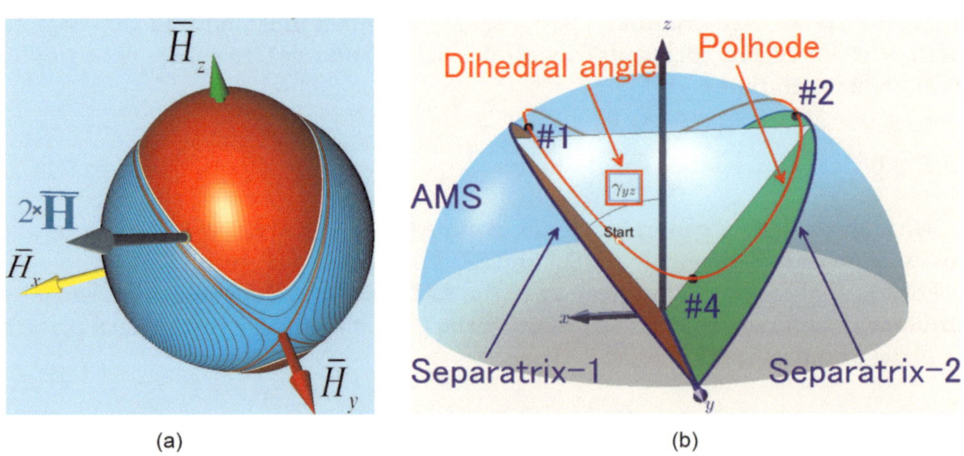

Figure 11.
(a) Plotted polhodes and separatrices on the AMS; (b) notations for separatrices.

Polhodes are seen on orthogonal projections as ellipses and hyperbolas. As for two separatrices, subdividing polhodes into four groups, they are seen as ellipses or as X-shaped two lines, illustrated in **Figure 11b** with the view from the intermediate axis y.

Dihedral angle for the separatrix can be determined, using the following Eq. (31):

$$\gamma_{yz} = \arctan \frac{\overline{H}_x}{\overline{H}_z}, \quad \text{where} \quad \frac{\overline{H}_x}{\overline{H}_z} = \sqrt{\frac{I_{xx}\left(I_{zz} - I_{yy}\right)}{I_{zz}\left(I_{yy} - I_{xx}\right)}} \tag{31}$$

In the literature, it is a common approach to present Poinsot's and Binet's constructions separately. However, as we wish to extract maximum advantages from both methods, we produce a co-centered combination of the conjugated virtual reality worlds (see **Figure 12a** and **12b**), corresponding to these two graphical methods and

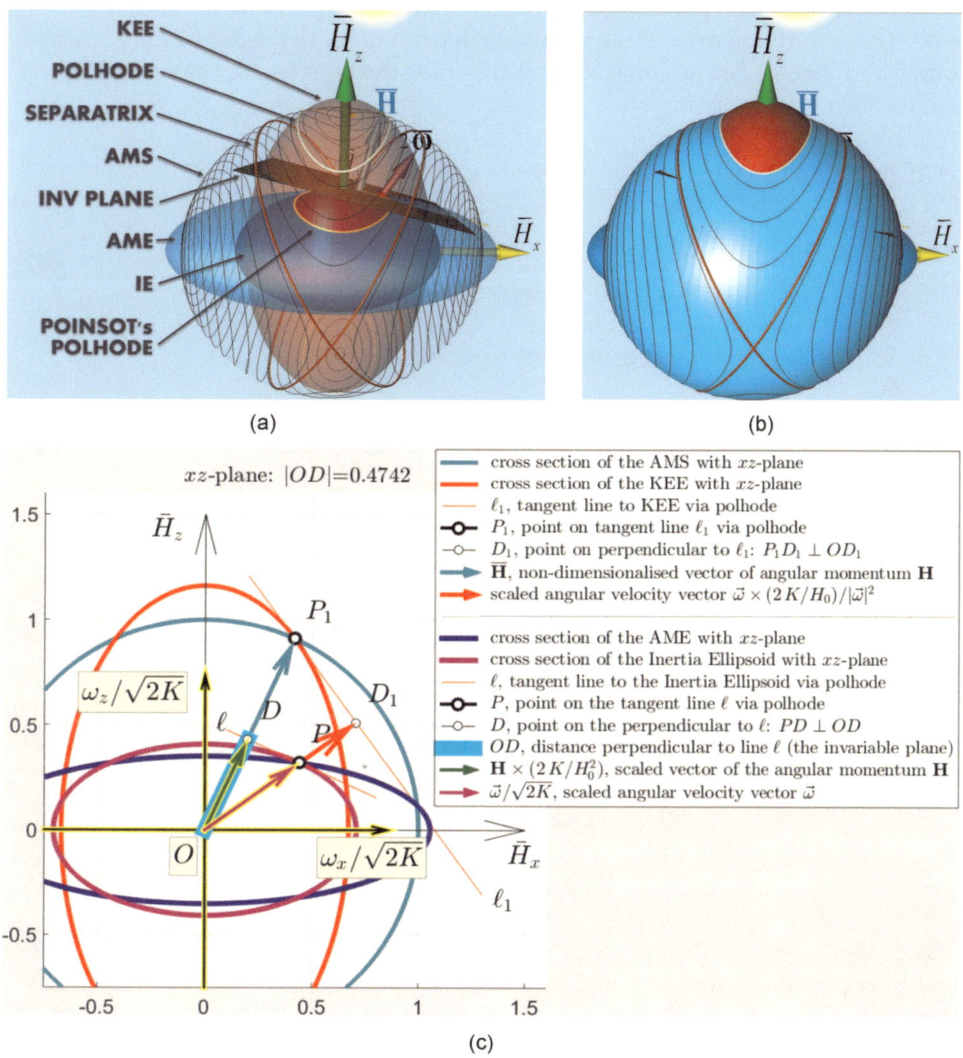

Figure 12.
Combined Poinsot's and Binet's constructions: (a) layout for the conjugated VR worlds with semi-transparent KEE, AMS, AME; (b) obscure KEE, AMS; (c) xz-cross section.

constructed for the system with $[I] = [2, \;\; 4, \;\; 6]$ and $[\omega_0] = [0.1, \;\; 1, \;\; 1]$. In **Figure 12c**, we also show the xz-cross section of these worlds. It is interesting to visually observe orthogonal directions of the semi-major axes in IE and KEE and also relative scaling of the vectors of angular velocity and angular momentum in two constructions. With the scaling adopted, the Poinsot's construction is at the lower level, compared to the Poinsot's construction, and, figuratively speaking, is "under the bonnet" in the combined methods setup.

5. Analysis of the unstable flipping motion

5.1 Calculation of the period of the tumbling and flipping motions

There are analytical expressions for the period of the periodic motion of the torque-free rotor; however, their notations depend upon the order of the principal moments of inertia. Let us consider first them for the $I_{xx} < I_{yy} < I_{zz}$ case, which, in turn, has two sub-cases:

- If $H^2 > 2K_0 I_{yy}$, which is equivalent to $a_y < 1$, then

$$T = 4\,\mathcal{K}\sqrt{\frac{I_{xx} I_{yy} I_{zz}}{(I_{zz} - I_{yy})(H^2 - 2K_0 I_{xx})}} \tag{32}$$

- If $H^2 < 2K_0 I_{yy}$, which is equivalent to $a_y > 1$, then

$$T = 4\,\mathcal{K}\sqrt{\frac{I_{xx} I_{yy} I_{zz}}{(I_{xx} - I_{yy})(H^2 - 2K_0 I_{zz})}} \tag{33}$$

where $\mathcal{K} = \mathcal{K}(k)$ is complete elliptic integral of the first kind:

$$\mathcal{K}(k) = \int_0^1 \frac{ds}{\sqrt{(1-s^2)(1-k^2 s^2)}} = \int_0^{\pi/2} \frac{du}{\sqrt{1-k^2 \sin^2 u}} \tag{34}$$

being a function of the parameter k $(-1 \le k \le 1)$, often called "elliptical modulus" and in the current task formulation is to be calculated as follows:

$$k = \sqrt{\frac{(I_{yy} - I_{xx})(2K_0 I_{zz} - H^2)}{(I_{xx} - I_{yy})(H^2 - 2K_0 I_{xx})}} \tag{35}$$

The elliptic integral Eq. (34) can be calculated, using various on-line calculators, sometimes providing amazing accuracy of up to 50 digits, if needed [24]. In many other cases, where this high precision is not a requirement, elliptic integral $\mathcal{K}$ can be calculated, using MATLAB® and Wolfram MATHEMATICA®; however, a special care should be given to the specifications of the relevant specialized functions to be employed. For example, $\mathcal{K}$ can be calculated, using MATLAB command `ellipke` or `ellipticK` command [25]; however, instead of k to be used in KEISAN calculator, k^2

(i.e., the squared value of k) should be provided as argument in these commands in MATLAB: this is due to the specific definition of these commands in MATLAB, presented for completeness below:

$$\mathcal{K}(m)|_{MATLAB} = \int_0^{\pi/2} \frac{d\theta}{\sqrt{1 - m \ \sin^2\theta}}, \tag{36}$$

Note that the elliptical modulus k or, in other terminology, the modular angle α are related to the parameter m as follows:

$$m = k^2 = \sin 2\alpha \tag{37}$$

In view of the different formats for the arguments in elliptic integral calculation, a series of simple verification tests may be advisable. Reconciliation of the specification differences is illustrated in **Figure 13** and the user can reproduce conforming results, using the following:

- KEISAN, by entering test input of k=0.5 and requesting 22-digit accuracy: **1.685750354812596**042871
- MATLAB: by calling any of the commands `ellipticK(0.25)` or `ellipke(0.25)`, where $0.25=0.5^2$ (with switch to the "long format"): **1.685750354812596**

In our case, we have high precision calculation of the period programmed in Java and embedded in the VR, enabling to get this information interactively straight in the virtual reality environment and display it in real time with any inertial morphings, applied to the conjugated concurrent configurations of the system.

The VR interface system, programmed by authors, is illustrated in **Figure 14**.

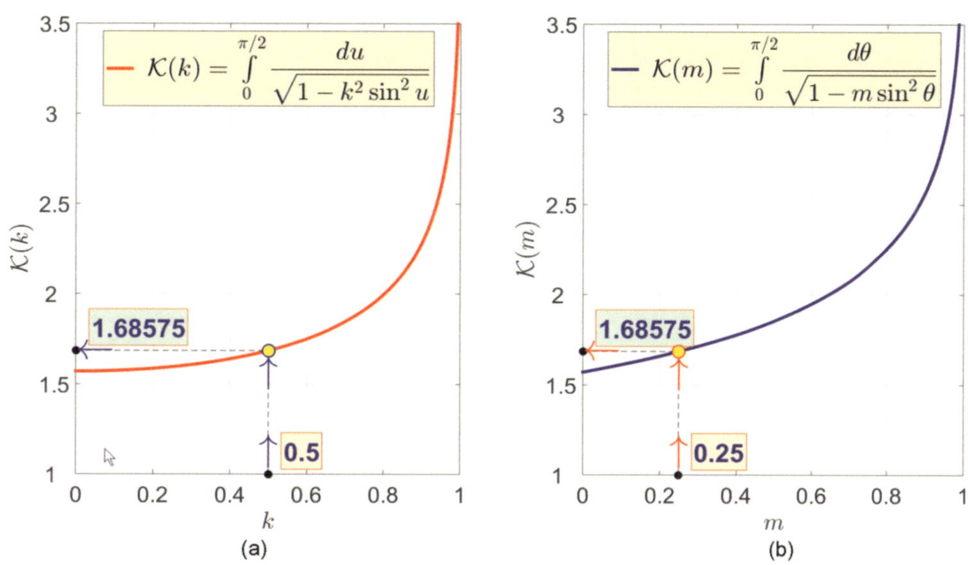

Figure 13.
Calculation of the complete elliptic integral of the first kind for the illustration example: (a) KEISAN definition with "k" argument; (b) MATLAB definition with "m" argument.

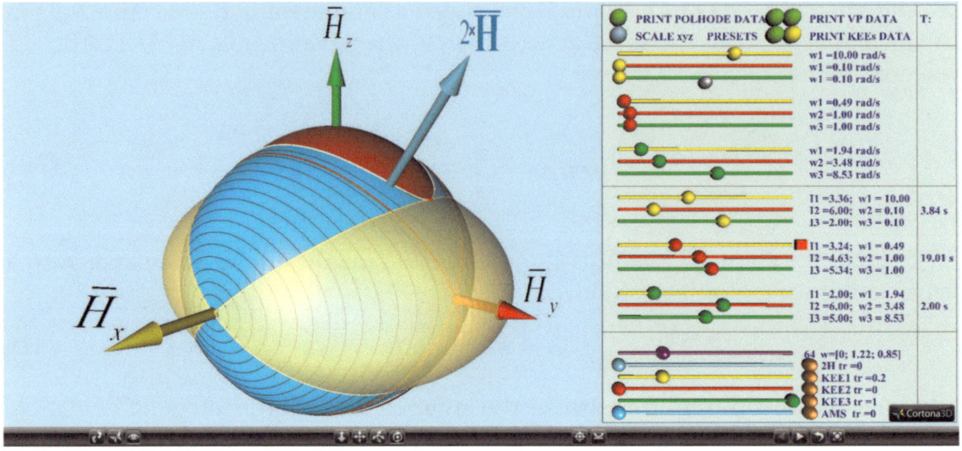

Figure 14.
Two conjugated configurations of the spacecraft, corresponding to the red and yellow kinetic energy ellipsoids KEEs.

Figure 14 displays two initial configurations of the spacecraft, using AMS and KEEs. The first, corresponding to the red KEE, has the following characteristics: I_{xx} = 3.24, ω_{x0} = 0.49, I_{yy} = 4.63, ω_{y0} = 1, I_{zz} = 5.34, ω_{z0} = 1. Instead of showing traditional vector of the angular momentum $\overline{\mathbf{H}}$, for visualization purposes, a magnified vector $2\overline{\mathbf{H}}$ is shown with the light blue color. Magnification should be applied, otherwise vector $\overline{\mathbf{H}}$, without magnification, would not be seen, as would be completely residing inside the AMS.

Parameters of the first configuration of the system are in compliance with $H^2 > 2K_0 I_{yy}$ condition.. The period of its tumbling motion, therefore, can be calculated, using Eq. (32), and is equal to T=19.01 s. Another configuration of the spacecraft, corresponding to the yellow KEE (shown as semi-transparent surface), has the following characteristics: I_{xx} =3.36, ω_{x0} = 10, I_{yy} =6, ω_{y0} = 0.1, I_{zz} = 2, ω_{z0} = 0.1. This configuration is in compliance with $H^2 < 2K_0 I_{yy}$ condition; therefore, for the calculation of the period of the motion, Eq. (34) should be used. Its application results in T=3.84 s. It may be of interest to notice that both KEEs have no intersections, but have four touching points, one of which is shown with the red dot, corresponding to the common set of the angular velocity components: $[\omega]$ = [0, 1.22, 0.85].

In our case, we have high precision calculation of the period programmed in Java and embedded in the VR, enabling to get this information interactively straight in the virtual reality environment and display it in real time with any inertial morphings, applied to the conjugated concurrent configurations of the system.

5.2 Differential equations of the torque-free spacecraft with Inertial Morphing

In order to be able to simulate the spacecraft with IM capabilities, the Euler's equations should be expanded, as follows, to allow change of the inertial properties of the system:

$$\begin{bmatrix} I_{xx} & 0 & 0 & 0 & 0 & 0 \\ 0 & I_{yy} & 0 & 0 & 0 & 0 \\ 0 & 0 & I_{zz} & 0 & 0 & 0 \\ 0 & 0 & 0 & \sin\theta\ \sin\phi & \cos\phi & 0 \\ 0 & 0 & 0 & \sin\theta\ \cos\phi & -\sin\phi & 0 \\ 0 & 0 & 0 & \cos\theta & 0 & 1 \end{bmatrix} \begin{Bmatrix} \dot{\omega}_x \\ \dot{\omega}_y \\ \dot{\omega}_z \\ \dot{\psi} \\ \dot{\theta} \\ \dot{\phi} \end{Bmatrix} = \begin{Bmatrix} (I_{yy} - I_{zz})\ \omega_y\omega_z - \dot{I}_{xx}\ \omega_x \\ (I_{zz} - I_{xx})\ \omega_z\omega_x - \dot{I}_{yy}\ \omega_y \\ (I_{xx} - I_{yy})\ \omega_x\omega_y - \dot{I}_{zz}\ \omega_z \\ \omega_x \\ \omega_y \\ \omega_z \end{Bmatrix}$$

where ψ, θ, and ϕ are 313 Euler angles [26].

5.3 Application of IM for stopping flipping motions or their initiation

In our earlier publications [9, 11], we suggested a method of switching between stable and unstable motions of the spinning systems. Unstable motion is possible only when the predominant rotation of the system is about its intermediate axis. Therefore, for the flipping systems, in order to stop flips, we suggested to use a method of deliberate change of the system's moments of inertia in a way to transform the axis of the predominant spin into the minimum or maximum inertia axis. Therefore, for stopping the flips, if they are not desired types of motion, there could be two strategies, illustrated in **Figure 15**.

In a similar way, the flipping motion can be initiated on a system, being initially in a regular stable spin. This could be only possible if the axis of the predominant rotation is initiated about the minimum or maximum principal axis. For initiation of the spin, when needed, IM can be applied to transform the axis of predominant rotation into the intermediate axis.

Pre-planning of the IM maneuver can be done with the VR interface, which has numerous features in one. It enables to include Poinsot's and Binet's constructions, display several designs concurrently, interactively review them in real time, and select the solution, satisfying key requirements. The starting configuration is set with the red sliders, evoking calculation of the periods of the motions, visualization of the IEs, AMEs, KEEs, AMS, patterns of the motion of the angular momentum and angular velocity vectors, polhodes, herpolhodes, separatrices, alternative separatrices, etc. Two additional conjugate configurations are also available for activation, allowing to foresee attitude motion of the system after the application of various inertial morphings.

One of the interface windows is displayed in **Figure 16**, where just for the illustration purposes, the KEEs for the conjugated variants are presented as semi-transparent.

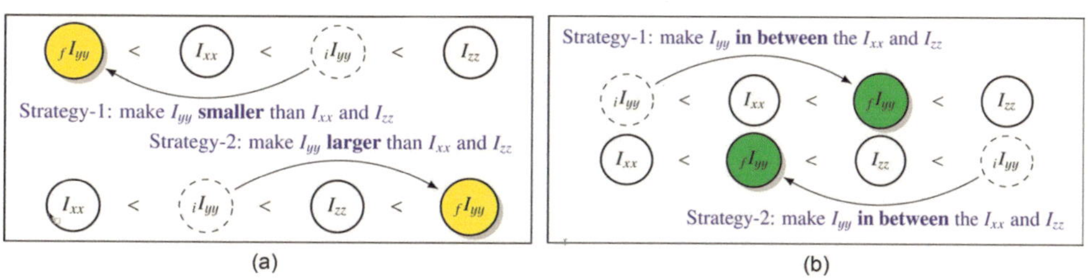

Figure 15.
Conceptual strategies (with multiple solutions) of: (a) stopping unstable flipping motion of the system about its intermediate axis of inertia, y-axis; (b) initiation of unstable flips about y-axis for two cases, when I_{yy} is initially the minimum or the maximum axis.

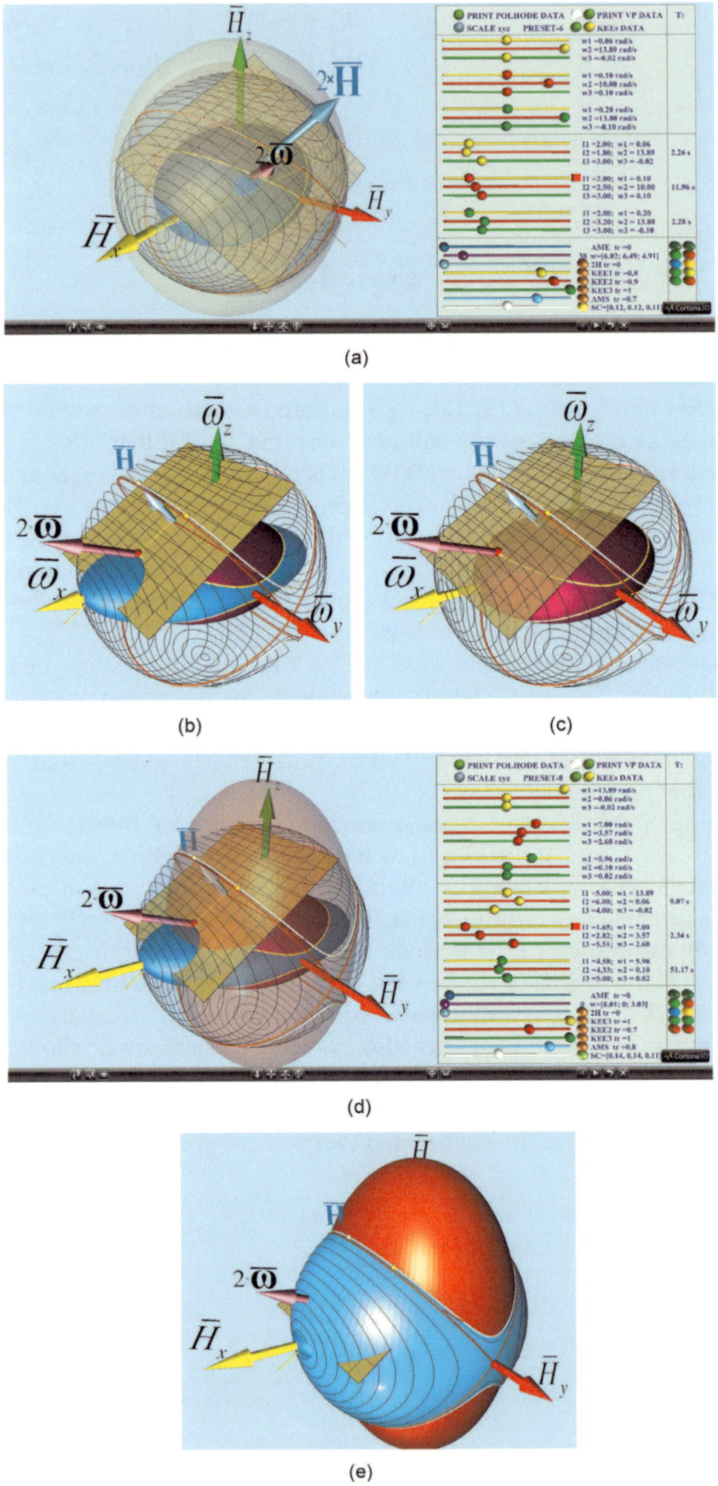

Figure 16.
Conjugated VR worlds, combining Poinsot's and Bine's constructions and provisions for three concurrent morphed variants of the system: (a) interface; (b-e) review stages.

The interface illustrates possibility to design a scenario of stopping the flipping motion. We consider the system which has the following initial conditions: $[I]$=[2, 2.5, 3] and $[\omega]$=[.1, 10, .1], corresponding to the flipping motion with the period of T=12 s.

There could be multiple solutions to the stabilization of the spacecraft. As illustration of the IM, we firstly use strategy-1 in **Figure 15a** and apply at the instant t=12.4 s the only one single prompt morphing, enabling change of the ${}_iI_{yy}$ of the system to its new value of ${}_fI_{yy}$=3.2. The results of the numerical simulations, partially presented in **Figure 17a**, confirm that with the applied IM, the system was transferred into the almost regular spin about the y-axis. This simulation is in agreement with the Law of Conservation of angular momentum, requiring and achieved increase of the dominant spin rate as per the $({}_iI_{yy}/\,{}_fI_{yy})$ ration: $10 \times ({}_iI_{yy}/\,{}_fI_{yy}) = 10 \times (2.5/3.2) \approx 8$. Furthermore, as illustration of strategy-2 we apply only one single prompt morphing, enabling change of the ${}_iI_{yy}$ of the system to its new value of ${}_fI_{yy}$ = 1.8. The results of the numerical simulations, partially presented in **Figure 17b**, confirm that with the applied IM, the system was transferred into the almost regular spin about the y-axis. This simulation is in perfect agreement with the Law of Conservation of angular momentum, requiring and achieved increase of the dominant spin rate as per the $({}_iI_{yy}/\,{}_fI_{yy})$ ratio: $10 \times ({}_iI_{yy}/{}_fI_{yy}) = 10 \times (2.5/1.8) \approx 14$. With the first solution, the KEE of the system "shrinks" getting smaller than the co-centered AMS with only two small touching spots on the y-axis (see **Figure 17c**). With the second solution, the KEE "expands" to completely embrace the AMS, with two small touching spots on the y-axis (see **Figure 17d**).

Various IM maneuvers can be combined together. An example of the possible practical application for the spacecraft is presented in **Figure 18**, showing 180-degrees inversion of the spacecraft. This enables utilization of the same thruster for boost and for braking. After conceptual design of the sequence of the IMs, mainly performed in the body axes, the VR interface can be further used to observe the attitude maneuver of the spacecraft in the inertial coordinates.

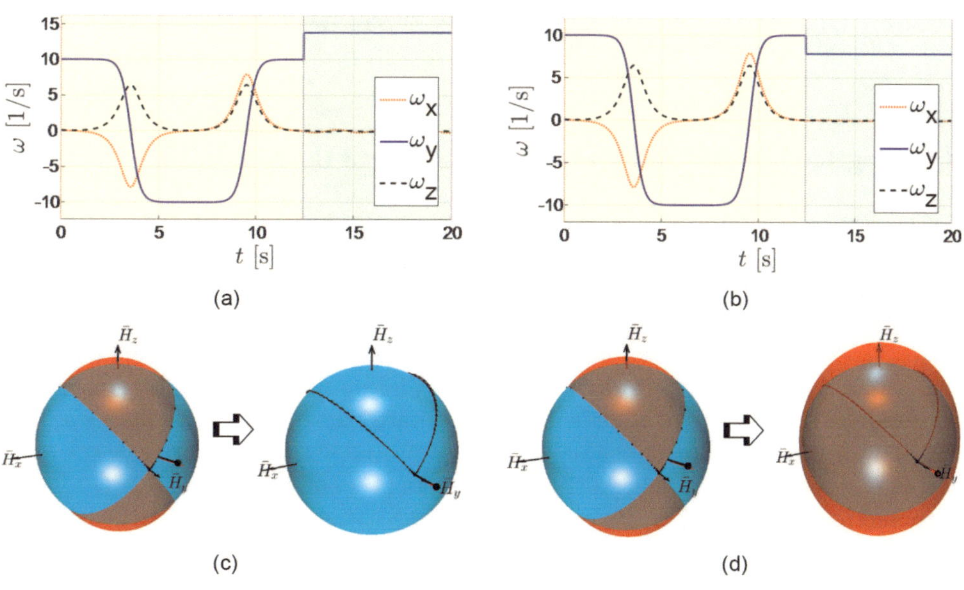

Figure 17.
Stopping flipping motions of the system with only one inertial morphing: (a) Strategy-1 with y becoming minimal axis of inertia; (b) Strategy-2 with y becoming maximum axis; (c) and (d) transformations of the KEEs, corresponding to Strategies-1 and 2.

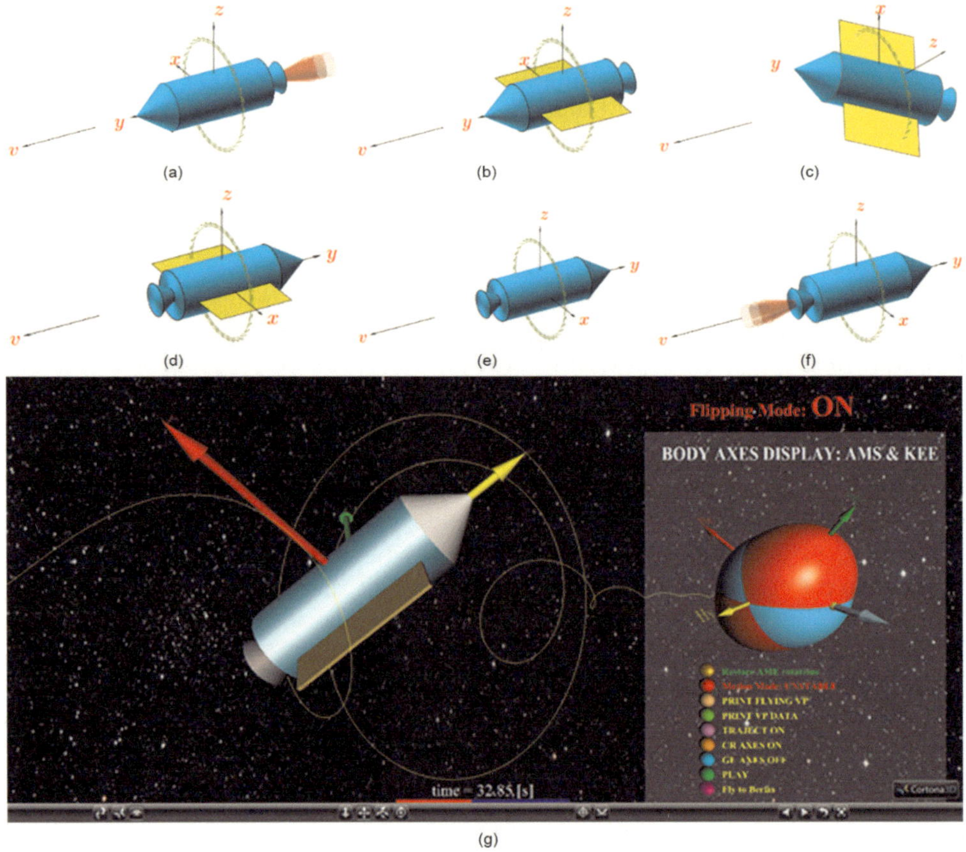

Figure 18.
Representation of the 180° inversion of the spacecraft, allowing use of the same thruster for boost and braking: (a–f) stages of the flight; (g) performing flip in VR.

5.4 Conjugated Virtual Reality assist de-tumbling of the spacecraft

For the system with three distinct principal area moments of inertia, there could be six different cases: $I_{xx} < I_{yy} < I_{zz}$; $I_{xx} < I_{zz} < I_{yy}$; $I_{yy} < I_{xx} < I_{zz}$; $I_{yy} < I_{zz} < I_{xx}$; $I_{zz} < I_{xx} < I_{yy}$; $I_{zz} < I_{yy} < I_{xx}$ and the format of the equations for polhodes and separatrices would depend upon the order of the moments of inertia. Instead of re-writing relevant equations for various cases, we will introduce a universal statement, employing the subscripts below, which distinguish minimal, intermediate, and maximum moments of inertia. In these notations, the dihedral angle γ of the separatrix, measured from the axis with maximum moment of inertia to the axis with minimal moment of inertia (note: order" max $\rightarrow$ min " is important), can be expressed as follows [21]:

$$\tan\gamma = \tan^{-1}\left(\frac{\overline{H}_{\min}}{\overline{H}_{\max}}\right), \quad \text{where} \quad \frac{\overline{H}_{\min}}{\overline{H}_{\max}} = \sqrt{\frac{I_{\min}\left(I_{\max} - I_{\text{intermed}}\right)}{I_{\max}\left(I_{\text{ntermed}} - I_{\min}\right)}} \tag{38}$$

Therefore, if at the particular instant of time, we have a tip of the vector $\overline{\mathbf{H}}$, marked as dot on the AMS with current coordinates $\overline{H}_{xC}$, $\overline{H}_{yC}$, and $\overline{H}_{zC}$, then we can use Eq. (38) to calculate $\tan\gamma$ and then can apply inertia morphing in such a way that the point would become on one of the separatrices of the morphed system. With instantly applied morphing, the dot would be "intercepted" by the separatrix and would start moving along the newly "constructed railway", instead of moving along its previous path. We called this method "insertion into separatrice" [23]. The easiest way to implement intercepts would be to assume them occurring when the dot is at one of the points of intersection of the dot's polhode with one of the *xy*, *yz*, or *xz* planes. To get the required paramters for the "insertion", we can assign any new values of the minimal and maximum moments of inertia ($I_{\min}$ and $I_{\max}$) and then use Eq. (38) to calculate the value of the intermediate moment of inertia, required for the transfer to occur:

$$I_{\text{intermed}} = I_{\min}\ I_{\max}\left(1 + \tan^2\gamma\right) / \left(I_{\min} + I_{\max}\tan^2\gamma\right) \tag{39}$$

The developed VR interface is very efficient in creating and "interviewing" various solutions and then selection of the most suitable maneuver scenario. Let us consider a particular numerical example. In **Figure 19a** we show the system with $[I]$ = [3.44, 6, 3], $[\omega_0]$ = [2.46, 1.44, 0.96], described with red KEE. Let us "intercept" the point, moving along its polhode, at the *xz*-plane and initiate its flipping motion along the "*y*" axis. For the point on its original polhode we have

$$\tan\gamma = H_{zC}/H_{yC} = 0.6313/0.7756 = 0.8139.$$

Therefore, further assuming that the hardware could ensure new morphed values of the minimum and maximum moments of inertia I_{yy} = 6; I_{zz} = 4, we can calculate the required new value of the intermediate moment of inertia I_{xx} = 5.0032. So, the set of the moments of inertia after the first IM is: $[I]_{1B}$ = [5, 6, 4]. However, the VR interface would immediately produce the same parameters, with quick interactive selection of the I_{yy} and I_{zz}. As an alternative to this solution, if the new values of the minimum and maximum moments of inertia would be selected as I_{yy} = 6; I_{zz} = 3, then we would get I_{xx} = 4.2904: $[I]_{1B}$ = [4.29, 6, 3]. Also, the third illustration solution would be selection of the following values for the first IM: $[I]_{1C}$ = [4.58, 4.33, 5]. Interestingly, cases *A* and *C* have the same separatrices, but have significant difference in execution time, which can be important for the planning of the maneuvre. In planning of the case A, red and yellow KEEs conjugates were employed; in planning of the case *C*, red and green conjugates were used. In the **Figure 19**, some of the surfaces were interactively made semi-transparent. In **Figure 19e** and **19f** we show the spacecraft VR world, merged together with the AMS+KEE VR world. For convenience, their motion can be synchronised or disconnected, enabling for the user to observe AMS-KEE in both body and inertial coordinate systems. In addition, change of the viewpoints enables observations of the spacecrat from its body axes or from inertial reference set.

To complete stabilization of the spacecraft, application of the second IM is needed: it should be applied at the moment when the vector $\overline{\mathbf{H}}$ is passing one of the '*y*' axis AMS poles. As illustration, we select new values after the second IM applied: $[I]_{A,B,C}$ = [6.5, 6, 4].

The VR planning maneuver interface, resultant time histories of the angular velocities for the two-morphing meneuvers (*A* and *C*), and the screen snapshots from the VR animations are shown in **Figure 19**.

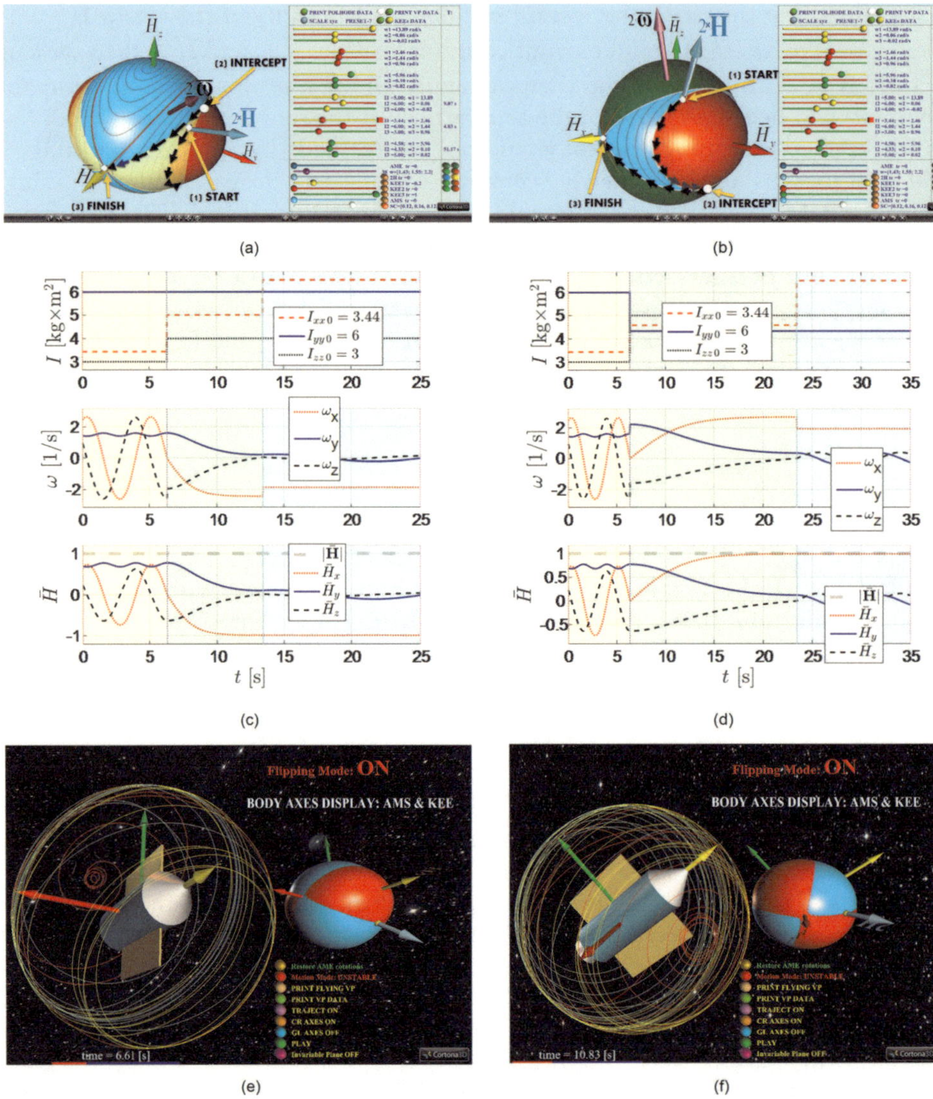

Figure 19.
Two competing variants of the de-tumbling of the spacecraft presented side-to-side: (a–b) maneuver planning interface; (c–d) simulation raw results; (e–f) animation in VR.

5.5 Conjugated Virtual Reality assist 90° inversion of the spacecraft

In this subsection we will illustrate the application of the IM for the 90-degrees spacecraft inversion with the sequence shown in **Figure 20**. For the planning, the VR system is used. As our system with conjugated VR worlds allows superposition of the various morphed configurations for the spacecraft, we can select two separatrices: the first—coming from the initial "destination" of the tip of the vector $\overline{\mathbf{H}}$ and the second —coming to the axis, to which the transfer of the rotation is to be passed. Their intersection constitutes the point, where instantaneous IM should be applied. In the interface in **Figure 21**, the first separatrix corresponds to the red KEE, and the second—to the yellow KEE.

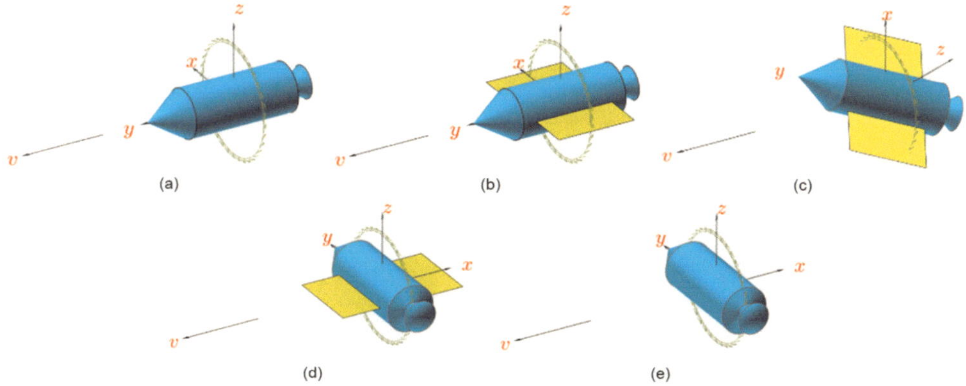

Figure 20.
Illustration of the 90-degrees inversion of the spacecraft, aiming to transfer its rotation from one axis to another axis at 90 degrees: (a–e) stages of the maneuver.

IM index	*t*, s	I_{xx}	I_{yy}	I_{zz}
0	0	3	2.8	5
1	10	3	3.1	5
2	18.25	4.5	3.1	5
3	25.15	5.5	3.1	5
f	35.14	5.5	3.1	5

Table 1.
IM parameters for the 90-degrees spacecraft inversion.

Remarkably, only three IMs would be required for implementation of such complex maneuver. One possible example of the parameters for the 90-degrees inversion is presented in **Table 1**.

Figure 21 presents results of the simulation of the selected case. It can be seen that the spacecraft, which initially was in stable spin about its longitudinal axis, after three IMs was transferred to the rotation about its lateral axis. This is confirmed not only with the plots of the components of the angular velocity in **Figure 21c**, but also with the plots of the Euler angles **Figure 21d**.

Using similar principle, it is possible to design cascaded maneuvers. For example, in [12] we designed the sequence, where the spacecraft performed all-axes inversions "parade". With the simplicity of the required control, involving only exiguous/paltry number of IM control actions, it is believed that the systems with IM could find wide application in autonomous small spacecraft.

6. Conclusions

This book chapter is further exploring application of the "inertial morphing" control principles and is dedicated to the design of the integrated VR platform, enabling design, test, optimization, and visualization of various missions for the autonomous spacecraft.

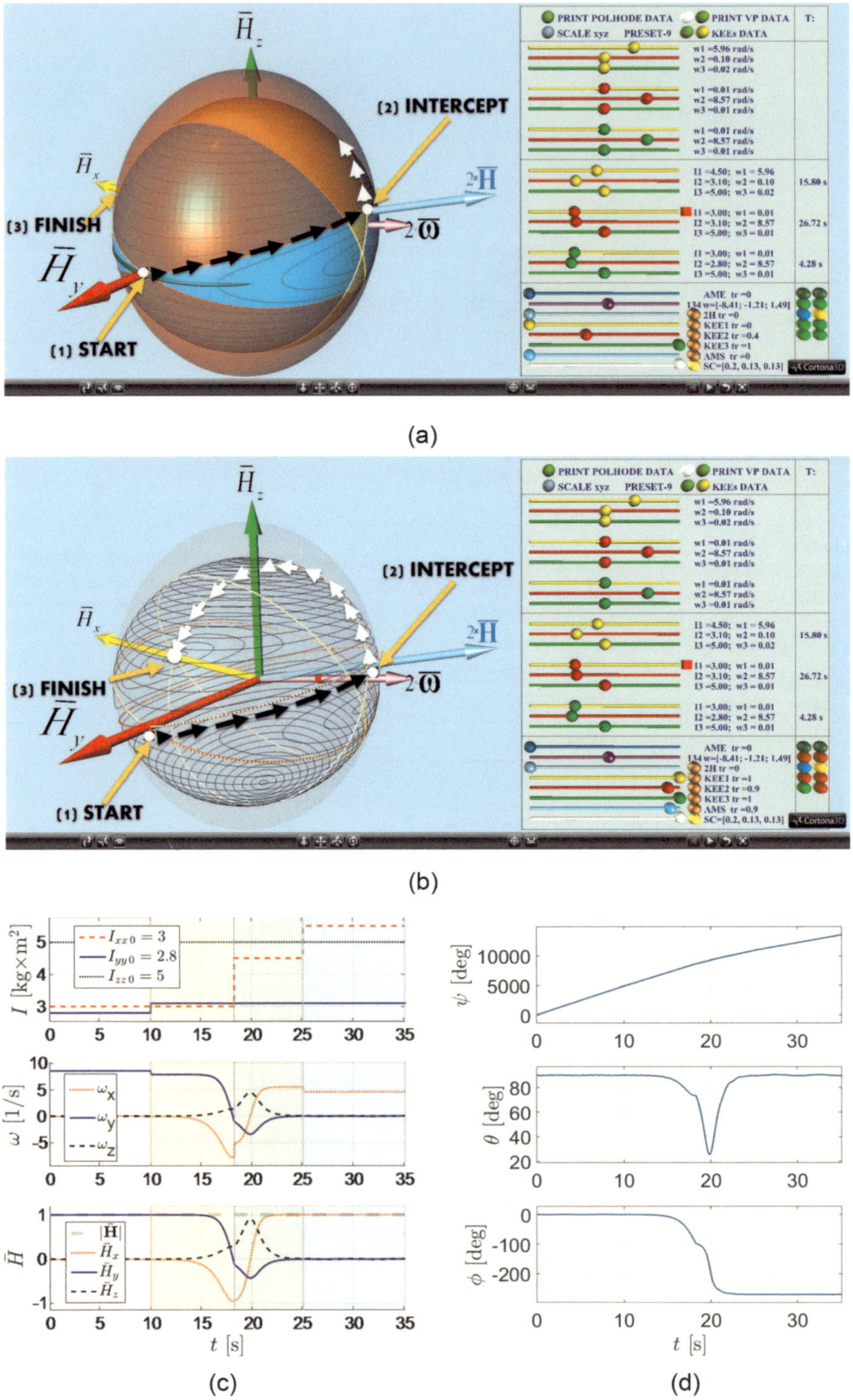

Figure 21.

Illustration of the utilization of the conjugated VR worlds for planning spacecraft maneuver: (a) selection of the "separatrix to separatrix" trajectory; (b) making AMS and KEEs highly transparent for the visualization of the transfer; (c) time history of components of the angular velocity; (d) time history of 313 Euler angles.

The key feature of this work is introduction of several VR worlds for effective evolutionary visualization of the attitude dynamics of the spacecraft. The demonstrated set of the conjugated VR worlds include the following: "Inertial VR World", "Body-Axes Non-Inertial VR World", "Inertial Geometric Interpretation VR Worlds", "Body-Axes Non-Inertial Geometric Interpretation VR World". This integrated virtual reality environment of several conjugated VR worlds proved to be an efficient tool for the design of the autonomous spacecraft missions with specific requirements of re-orientation, acrobatic attitude maneuvering, for simulation of the competing scenarios, their optimization, and animated visualization. Most significantly, for the inertially morphed systems, it enables design of the astonishingly simple, limited number discrete control inputs. For example, it enables design of the inversion of the spacecraft with two only instantaneous control adjustments. And the VR environment also enables design of the spacecraft de-tumbling, as well as 90-degrees inversion of the spacecraft with only three instantaneous control adjustments.

Author details

Pavel M. Trivailo[1]* and Hirohisa Kojima[2]

1 RMIT University, Melbourne, Australia

2 Tokyo Metropolitan University, Tokyo, Japan

*Address all correspondence to: pavel.trivailo@rmit.edu.au

References

[1] Fuchs P, Moreau G, Guitton P, editors. Virtual Reality: Concepts and Technologies. London, UK: CRC Press; 2011

[2] Baughman A. Evaluation of Virtual and Hybrid Reality Systems for Astronaut Training. Colorado: University of Colorado Boulder; 2020

[3] Twombly A et al. The virtual GloveboX (VGX): A Semi-immersive Virtual Environment for Training Astronauts in Life Science Experiments. Systemics, Cybernetics and Informatics. 2006;**2**(3):30-34

[4] Seymour NE. VR to OR: A review of the evidence that virtual reality simulation improves operating room performance. World Journal of Surgery. 2008;**32**(2):182-188

[5] Garcia AD, Schlueter J, Paddock E. Training astronauts using hardware-in-the loop simulations and Virtual Reality. In: AIAA Scitech. Orlando, FL, USA; 2020

[6] Aoki H, Oman CM, Buckland DA, Natapoff A. Desktop-VR system for preflight 3D navigation training. Acta Astronautica. 2008;**63**(7–10):841-847. DOI: 10.1016/j.actaastro.2007.11.001

[7] Somavarapu DH, Guzzetti D. Toward immersive spacecraft trajectory design: Mapping user drawings to natural periodic orbits. Acta Astronautica. 2021; **184**:208-221. DOI: 10.1016/j.actaastro. 2021.04.004

[8] Piechowski S, Pustowalow W, Arz M, Rittweger J, Mulder E, Wolf OT, et al. Virtual reality as training aid for manual spacecraft docking. Acta Astronautica. 2020;**177**:731-736

[9] Trivailo PM, Kojima H. Utilisation of the "Dzhanibekov's Effect" for the possible future space missions. In: Proceedings of the Joint Conference: 31st ISTS, 26th ISSFD & 8th NSAT. Matsuyama, Japan; 2017. pp. 1-10

[10] Trivailo PM, Kojima H. Re-discovering "Dzhanibekov's Effect" using non-linear dynamics and virtual reality. In: Proceedings of the 68th International Astronautical Congress (IAC). Adelaide, Australia; 2017. p. 14

[11] Trivailo PM, Kojima H. Inertial morphing as a novel concept in attitude control and design of variable agility acrobatic autonomous spacecraft. In: Dai L, Jazar RN, editors. Nonlinear Approaches in Engineering Application. Cham, Switzerland: Springer; 2022. pp. 119-244. DOI: 10.1007/978-3-030-82719-9_5

[12] Trivailo PM, Kojima H. Enhancement of the spinning spacecraft attitude dynamics capabilities using inertial morphing. Royal Aeronautical Society The Aeronautical Journal. 2019; **12**:1-34. DOI: 10.1017/aer.2019.145

[13] Trivailo PM, Rittweger A, Theil S. Utilisation of the controllable inertial morphing for providing spacecraft with acrobatic attitude capabilities. In: Preprints of the 21st IFAC World Congress (Virtual). Berlin, Germany; 2020. pp. 15105-15110

[14] Poinsot, Théorie nouvelle de la rotation des corps, Paris, 1834

[15] List of the 72 names on the Eiffel Tower. https://en.wikipedia.org/wiki/List_of_the_72_names_on_the_Eiffel_Tower [Accessed: November 18, 2022]

[16] Eiffel Tower. Image Courtesy Benh LIEU SONG - File:Tour Eiffel Wikimedia

Commons.jpg, CC BY-SA 3.0, https://commons.wikimedia.org/w/index.php?curid=7677663 [Accessed: November 18, 2022]

[17] Poinsot' plaque. Image Courtesy Juanedc from Zaragoza, España - Los nombres de la Torre Eiffel, CC BY 2.0, https://commons.wikimedia.org/w/index.php?curid=68339501 [Accessed: November 18, 2022]

[18] Goldstein H, Poole C, Safko J. Classical Mechanics. Reading, MA: Addison-Wesley; 2000

[19] Whittaker ET. A Treatise on the Analytical Dynamics of Particles and Rigid Bodies. 2nd ed. Cambridge: University Press; 1917. p. 432

[20] Spivak M. Physics for Mathematicians: Mechanics I. Houston, TX, USA: Publish or Perish Publisher; 2010. p. 733

[21] Trivailo PM, Kojima H. Augmented control of inversion of the spinning spacecraft, using inertial morphing, paper IAC-18,C2,3,5,x45333. In: Proceedings of the 69th International Astronautical Congress. Bremen, Germany; 2018. p. 16

[22] Trivailo PM, Kojima H. Discovering Method of Control of the "Dzhanibekov's Effect" and Proposing its Applications for the Possible Future Space Missions. – Transaction of JSASS (The Japan Society for Aeronautical and Space Sciences). Aerospace Technology Japan. 2019;**17**(1):72-81. DOI: 10.2322/tastj.17.72

[23] Trivailo PM, Kojima H. Enhancement of the spacecraft attitude dynamics capabilities via combination of the inertial morphing and reaction wheels (Keynote Paper and Presentation). In: Transactions of the 18th Australian International Aerospace Congress (AIAC2019), incorporating 27th International Symposium on Space Flight Dynamics (ISSFD). Melbourne, VIC, Australia: Engineers Australia, Royal Aeronautical Society. 2019. pp. 941-970. DOI: 10.3316/informit.9781925627213

[24] KEISAN. Complete elliptic integral of the 1st kind *K(k)* Calculator, English version: https://keisan.casio.com/exec/system/1180573451 [Accessed: November 18, 2022]

[25] `ellipticK`: Complete elliptic integral of the first kind, https://au.mathworks.com/help/symbolic/sym.elliptick.html [Accessed: November 18, 2022]

[26] Rimrott FPJ. Introductory Attitude Dynamics. Mechanical Engineering Series Book. New York: Springer-Verlag; 1989. p. 383